Transformations:
A Mathematical Approach

Fundamental Concepts

Edited by

Carlos Polanco

Department of Mathematics, Faculty of Sciences
Universidad Nacional Autónoma de México, México

General:

1. Any dispute or claim arising out of or in connection with this License Agreement or the Work (including non-contractual disputes or claims) will be governed by and construed in accordance with the laws of the U.A.E. as applied in the Emirate of Dubai. Each party agrees that the courts of the Emirate of Dubai shall have exclusive jurisdiction to settle any dispute or claim arising out of or in connection with this License Agreement or the Work (including non-contractual disputes or claims).
2. Your rights under this License Agreement will automatically terminate without notice and without the need for a court order if at any point you breach any terms of this License Agreement. In no event will any delay or failure by Bentham Science Publishers in enforcing your compliance with this License Agreement constitute a waiver of any of its rights.
3. You acknowledge that you have read this License Agreement, and agree to be bound by its terms and conditions. To the extent that any other terms and conditions presented on any website of Bentham Science Publishers conflict with, or are inconsistent with, the terms and conditions set out in this License Agreement, you acknowledge that the terms and conditions set out in this License Agreement shall prevail.

Bentham Science Publishers Ltd.
Executive Suite Y - 2
PO Box 7917, Saif Zone
Sharjah, U.A.E.
Email: subscriptions@benthamscience.org

BENTHAM SCIENCE

CONTENTS

The essence of mathematics lies in its freedom

–Georg Cantor

FOREWORD

In this book, Dr. Carlos Polanco overcomes the usual gap between mathematicians and application users, describing with several examples the mathematical operator named transformation. This book can be used as auxiliary for students interested in this field as well as a reference book for seasoned Researcher.

Jorge Alberto Castañón González
Department of Critical Care Medicine
Hospital Juárez de México
México City
México

PREFACE

A "transformation", in its broadest sense, is an operator that relates two quantities. Its use permeates many every day artistic and scientific activities. An example of this would be its use in the manufacturing of tools to build and transform objects in different artistic expressions such as painting, sculpture, and theatre to determine the correct distance and size of things. In the scientific field, they are used in different disciplines such as Optics and Physics and more recently in Genomics and Bioinformatics. Due to the role transformations play in a broad spectrum of human creation, it is surprising to realize that few publications have been allocated to its study, not only in its mathematical context but also in its application to other disciplines. The reader will find in this book a study of this mathematical operator described by a large number of carefully selected examples, the reading and application, which can be easily understood without any rigorous mathematical knowledge. The first part introduces the affine transformations as mathematical operators, their properties, and their definition. The second part describes the differences between linear and nonlinear transformations. And the third part applies transformations in Acoustics, Actuary, Bioinformatics, Calculus, Cybernetics, Epidemiology, Genetics, Optics, Physics, Probability, and Vector Analysis. The author hopes the reader interested in Transformation theory finds useful the material presented here and that those who start studying this field, find this information motivating.

CONSENT FOR PUBLICATION

Not applicable.

CONFLICT OF INTEREST

The authors declare no conflict of interest, financial or otherwise.

ACKNOWLEDGEMENTS

Foremost, I would like to express my sincere gratitude to all contributors from the Faculty of Sciences at Universidad Nacional Autónoma de México for their collaboration in the preparation of this book.

Carlos Polanco
Department of Mathematics, Faculty of Sciences
Universidad Nacional Autónoma de México
México

List of Contributors

Alma Fernanda Sánchez Guerrero Department of Mathematics, Faculty of Sciences, Universidad Nacional Autónoma de México, México

Carlos Ignacio Herrera Nolasco Department of Mathematics Faculty of Sciences Universidad Nacional Autónoma de México, México

Carlos Polanco Department of Mathematics, Faculty of Sciences, Universidad Nacional Autónoma de México, México

Dánae Itzel Álvarez Figueroa Department of Actuary, Faculty of Sciences, Universidad Nacional Autónoma de México, México

Introduction

Carlos Polanco[*]

Faculty of Sciences, Universidad Nacional Autónoma de México, México

Abstract: Operator algebra is a branch of mathematics that can be applied to several fields of science as operators connecting vector spaces to interpret data from different algebras. Operator algebra can also be used for its axiomized properties. Both approaches are explored in this book.

Keywords: Affine transformation, Linear transformation, Nonlinear transformation, Transformation.

CONSIDERATIONS

This book shows the role of affine transformations in science introducing different scientific issues; it highlights the importance of mapping as a support of an equation. Although mapping Carrell [2005] Ch. 6 in a general sense, connects a vector space with another vector space, it is necessary to discuss some of the reasons why these operators are used as in most cases, they are nonlinear mapping. One of the reasons why mapping is necessary is because the algebra used on a vector space does not allow a practical approach to the solution of a problem. In this case, mapping will translate the problem to another vector space, where an alternative algebra reduces the complexity of the problem and makes it possible to find a solution. For instance, a problem involving multiple variables solved by a system of partial differential equations is likely to have no possible solution under that approach, however, if a matrix system is used, it is most likely to have a quick solution. In many cases, this degree of complexity can be interpreted with a nonlinear mapping. The importance of mapping is so that this book is based on it.

[*]**Corresponding author Carlos Polanco:** Faculty of Sciences, Universidad Nacional Autónoma de México, México City, México; Tel: +01 55 5622 4858; Fax: +01 5556 4859; E-mail: polanco@unam.mx

CHAPTER 1

Affine Transformations

Carlos Polanco[*]

Faculty of Sciences, Universidad Nacional Autónoma de México, México

   Abstract: This chapter focuses on the characterization of a mathematical operator called *transformation*; it describes the different types of transformations, their basic properties, restrictions, and extensions. The aspects here reviewed will be described in detail in the next chapters.

Keywords: Affine transformation, Automorphism, Bijective transformation, Composition transformation, Continuous function, Endomorphism, Function, Homothetic transformation, Injective transformation, Inverse transformation, Isomorphism, Linear space, Linear transformation, Nonlinear transformation, Reflection transformation, Rotation transformation, Scaling transformation, Shear transformation, Surjective transformation, Transformation, Translation transformation.

1.1. AFFINE TRANSFORMATION

Definition 1.1. An **affine** transformation [George, 1982; Parkhomenk and Parkhomenko, 1965] $T: A \rightarrow B, M(a) + b$ is a rule that assigns to every element $x \in A$ a unique element $y = T(x) \in B$, where $M \in A$ is a linear transformation ($\neq 0$), and $b \in B$ is a vector.

An affine transformation is classified as **linear transformation** if it meets (Def. 1.1) otherwise the affine transformation is a **nonlinear transformation**.

Definition 1.2. Composition transformation is the pointwise application of one transformation to the result of another to produce a third transformation. The composition transformation is not a commutative application. Be $T_1: A \rightarrow B$ and $T_2: B \rightarrow C$ two transformations. The **composition** will be $T_1 \circ T_2: A \rightarrow C = T_1(T_2)$ and its **inverse transformation** would be $T_1^{-1} \circ T_2^{-1}: C \rightarrow AT_1^{-1}(T_2)$.

Example 1.1. Let T_1, T_2 be nonlinear transformations $T_1: \mathbb{R}^2 \rightarrow \mathbb{R}^2$, $(x + 2, y + 2)$ and $T_2: \mathbb{R}^2 \rightarrow \mathbb{R}^2$, $(x - 2, y - 2)$. (i) Is $T_1 \circ T_2$ nonlinear? (ii) What about $T_2 \circ T_1$?

[*]**Corresponding author Carlos Polanco:** Faculty of Sciences, Universidad Nacional Autónoma de México, México City, México; Tel: +01 55 5622 4858; Fax: +01 5556 4859; E-mail: polanco@unam.mx

Solution 1.1. (i) Let $T_3 = T_1 \circ T_2 : T_1((x-2, y-2)) = (x,y)$. T_3 is linear. (ii) Let $T_4 = T_2 \circ T_1 : T_2((x+2, y+2)) = (x,y)$. T_4 is linear too.

Example 1.2. Be a nonlinear transformation $T_1 : \mathbb{R} \to \mathbb{R}$, x^2 and a linear transformation $T_2 : \mathbb{R} \to \mathbb{R}$, $x-2$. (i) Is $T_1 \circ T_2$ nonlinear? (ii) What about $T_2 \circ T_1$?

Solution 1.2. (i) Let $T_3 = T_1 \circ T_2 : T_1(x-2) = (x-2)^2 = x^2 - 4x + 4$. Since the first term is nonlinear, T_3 is nonlinear. (ii) Let $T_4 = T_2 \circ T_1 : T_2(x^2) = x^2 - 2$. Since the first term is nonlinear, T_4 is nonlinear.

Definition 1.3. Let T be a transformation whose domain is set A. T transformation is said to be **injective** provided that $\forall\ a, b \in A$, whenever $T(a) = T(b)$, then $a = b$; that is, $T(a) = f(b)$ implies $a = b$. Equivalently, if $a \neq b$, then $T(a) \neq T(b)$.

Example 1.3. Be a nonlinear transformation $T: \mathbb{R} \to \mathbb{R}$, x^2. Is T injective?

Solution 1.3. Let $T(-a) = T(a)$ but $-a \neq a$, therefore, T is not injective.

Example 1.4. Be a linear transformation $T: \mathbb{R} \to \mathbb{R}$, $x+3$. Is T injective?

Solution 1.4. Let $T(a) = T(b)$, since $a \neq b$, T is injective.

Definition 1.4. A transformation $T: A \to B$ with domain A and image B is surjective if for every $b \in B$ there exists at least one $a \in A$ with $T(x) = y$.

Example 1.5. Be a nonlinear transformation $T: \mathbb{R} \to \mathbb{R}$, x^2. Is T surjective?

Solution 1.5. If we take element $-1 \in \mathbb{R}$ in the image of T, there is not any element in the domain of T that $T(a) = -1$. Hence, T is not surjective.

Example 1.6. Be a linear transformation $T: \mathbb{R}^2 \to \mathbb{R}^2$, (x,y). Is T surjective?

Solution 1.6. $\forall\ T(x,y) \in \mathbb{R}^2$ in the image of T there exists at least one $(x,y) \in \mathbb{R}^2$ in the domain of T. Thus, T is surjective.

Definition 1.5. A **bijective** transformation $T: A \to B$ is an injective and surjective transformation of set A to set B.

Definition 1.6. Let T be a transformation whose domain is set A and whose image is set B. Then T is **invertible** if there exists a transformation T^{-1} with domain B and image A, with the property: $T(a) = b \Leftrightarrow T^{-1}(b) = a$. $T(a) = b \Leftrightarrow T^-(b) = a$.

Note 1.1. A **bijective linear transformation** is an **isomorphism**. A **linear transformation** on the same **linear space** is an endomorphism. If a **linear**

transformation is an **isomorphism** and an **endomorphism**, then it is an **automorphism**.

Example 1.7. Be an affine transformation $T: \mathbb{R} \to \mathbb{R}, \ e^x$. (i) Is T linear? (ii) Does T^{-1} exist?

Solution 1.7. (i) Let $T(x_1) = e^{x_1}, T(x_2) = e^{x_2}$, then $T(x_1) + T(x_2) = e^{x_1} + e^{x_2}$, but $T(x_1 + x_2) = e^{x_1}e^{x_2}$. Therefore T is not linear. (ii) If $T^{-1}(x) = \ln x \to T \circ T^{-1} = T(T^{-1}) = Id(x) = x$, then the inverse transformation is $T^{-1}(x) = \ln x$.

Functions are a particular type of transformations.

Definition 1.7. A function f meets $f(a)$ if $\forall \ a \in$ set A (domain of the function), corresponds a unique element of $b \in B$ where $b = f(a)$.

Example 1.8. Be function $f: \mathbb{R} \to \mathbb{R}, \ 3x$. (i) Is f surjective?, (ii) Is f injective?

Solution 1.8. (i) $\forall \ a \in A$ exists a unique element in set B and $f: \mathbb{R} \to \mathbb{R}$. So f is surjective. (ii) Let $f(a) = f(b)$ but $a \neq b$, therefore, f is injective.

Example 1.9. Be function $f: \mathbb{R} \to \mathbb{R}, \ x^2$. (i) Is f injective? (ii) Is f surjective? (iii) Is f bijective?

Solution 1.9. (i) $f(a) = f(-a)$ but $a \neq -a$, hence, f is not injective. (ii) Any element in $\mathbb{R}^-$ corresponds to an element in the image of f. Thus, f is not surjective. (iii) f is neither injective nor surjective so f is not bijective.

1.2. LINEAR TRANSFORMATIONS

A **linear transformation** $T: A \to B$ is a binary operation or "rule" that assigns to every element a of A a unique element $b = T(a) \in B$ [Carrell, 2005] Ch. 6.

Definition 1.8. Suppose A and B are **linear spaces** over field F. A transformation $T: A \to B$ is said to be **linear** if $\forall \ x, y \in A$ and $\forall \ \alpha \in F$ (1.1, 1.2) are met.

$$T(x + y) = T(x) + T(y) \tag{1.1}$$

$$\alpha T(x) = T(\alpha x) \tag{1.2}$$

Note 1.2. In Chapter 2 the reader will find this topic more elaborated.

Example 1.10. Show that $T: \mathbb{R} \to \mathbb{R}, \ mx$ is linear.

Solution 1.10. Let $x, y \in \mathbb{R}$ be the domain of T and $m \in \mathbb{R}$, then (i) $T(x + y) = m(x + y) = mx + my = T(x) + T(y)$; (ii) $\alpha T(x) = \alpha mx = T(\alpha x)$. So T is linear.

Example 1.11. Show that $T: \mathbb{R}^2 \to \mathbb{R}^2$, $(\alpha x, \beta y)$, where α, β are scalars, is linear.

Solution 1.11. Let (x_1, y_1), $(x_2, y_2) \in \mathbb{R}^2$ be the domain of T, then (i) $T(x_1 + x_2, y_1 + y_2) = [\alpha(x_1 + x_2), \beta(y_1 + y_2)] = (\alpha x_1 + \alpha x_2, \beta y_1 + \beta y_2) = (\alpha x_1, \beta y_1) + (\alpha x_2, \beta y_2) = T(x_1, y_1) + T(x_2, y_2)$; (ii) $\alpha T(x, y) = (\alpha x, \beta y) = T(\alpha x, \beta y)$. T is linear.

Example 1.12. Suppose V is the space of continuous real valued functions. Show that $T: \mathbb{V} \to \mathbb{R}$, $\int_a^b f(x)\,dx$ is linear [Carrell, 2005] Ch. 6.

Solution 1.12. Let $f, g \in V$, then (i) $T(f + g) = \int_a^b f(x) + g(x)\,dx = \int_a^b f(x)\,dx + \int_a^b g(x)\,dx = T(f) + T(g)$; (ii) $\alpha T(f) = \alpha \int_a^b f(x)\,dx = \int_a^b \alpha f(x)\,dx = T(\alpha f x)$. T is linear.

Example 1.13. Show that $T: \mathbb{R} \to \mathbb{R}$, $\sin x$ is nonlinear.

Solution 1.13. Let $x, y \in \mathbb{R}$ be the domain of T, then (i) $T(x + y) = \sin(x + y) = \sin x \cos y + \cos x \sin y \neq T(x) + T(y) = \sin x + \sin y$; (ii) $\alpha T(x) = \alpha \sin x \neq T(\alpha x) = \sin(\alpha x) = \sin \alpha \cos x + \cos \alpha \sin x =$. T is nonlinear.

Example 1.14. Show that $T: \mathbb{R} \to \mathbb{R}$, $mx + \beta$ is nonlinear.

Solution. 1.14. Let $x, y \in \mathbb{R}$ be the domain of T and $m, \beta \in \mathbb{R}$, then (i) $T(x + y) = m(x + y) = m(x + y) + \beta \neq T(x) + T(y) = mx + \beta + my + \beta = m(x + y) + 2\beta$; even if (ii) $\alpha T(x) = T(\alpha x)$. T is nonlinear.

Example 1.15. Show that $T: \mathbb{R} \to \mathbb{R}$, x^2 is nonlinear.

Solution 1.15. Let $x, y \in \mathbb{R}$ be the domain of T, then (i) $T(x + y) = (x + y)^2 = x^2 + y^2 + 2xy \neq T(x) + T(y) = x^2 + y^2$; (ii) $\alpha T(x) = \alpha x^2 \neq T(\alpha x) = \alpha^2 x^2$. T is nonlinear.

A linear transformation has three main sub-classifications: **rotation, reflection,** and **shear**.

1.2.1. Rotation Transformation

A **rotation** is a linear transformation where the figure is rotated around a fixed point called center of rotation. As a result, the distance from the center of rotation to any point on the figure is the same [Brannon, 2002; Nathan, 2009].

Example 1.16. Be transformation $T: \mathbb{R}^2 \to \mathbb{R}^2$, $(x\cos\theta - y\sin\theta, x\sin\theta + y\cos\theta)$, where θ is the rotation angle. Is T linear? (ii) Is T a rotation?

Solution 1.16. (i) Let (x_1, y_1), $(x_2, y_2) \in \mathbb{R}^2$ be the domain of T, then (a) $T(x_1, y_1) = (x_1\cos\theta - y_1\sin\theta, \ x_1\sin\theta + y_1\cos\theta)$ and $T(x_2, y_2) = (x_2\cos\theta - y_2\sin\theta, \ x_2\sin\theta + y_2\cos\theta) \Rightarrow T(x_1, y_1) + T(x_2, y_2) = [(x_1 + x_2)\cos\theta - (y_1 + y_2)\sin\theta, x_1 + x_2)\sin\theta + (y_1 + y_2)\cos\theta)] = T(x_1 + x_2, y_1 + y_2)$; (b) $\alpha T(x, y) = \alpha(x\cos\theta - y\sin\theta, x\sin\theta + y\cos\theta) = T(\alpha x, \alpha y)$. T is linear. (ii) Let (x, y) be any point in the domain of T, then $T(x, y) = (x\cos\theta - y\sin\theta, x\sin\theta + y\cos\theta)$. Taking the point $(1,1) \in C^1$ (unit circle) and the angle $\theta = \frac{\pi}{2} \Rightarrow T(1,1) = (-1,1)$ and $T(-1,1) = (-1,-1)$. Note that $\{(-1,1), (-1,-1), (1,1)\} \subset C^1$. T is a rotation.

Definition 1.9. The linear transformation $T: \mathbb{R}^2 \to \mathbb{R}^2$, $(x\alpha\cos\theta - y\alpha\sin\theta, x\alpha\sin\theta + y\alpha\cos\theta)$, where θ is the rotation angle and α is the distance from the rotation center, is a generalization (Example 1.16).

1.2.2. Reflection Transformations

In this type of transformation each point is at the same distance from the central line. A **reflection** has the same size as the original figure.

Example 1.17. Be transformation $T: \mathbb{R}^2 \to \mathbb{R}^2$, $(x, -y)$; (i) Is T linear? (ii) Is T a reflection?

Solution 1.17. (i) Let (x_1, y_1), $(x_2, y_2) \in \mathbb{R}^2$ be the domain of T, then (a) $T(x_1, y_1) = (x_1, -y_1)$, and $T(x_2, y_2) = (x_2, -y_2) \Rightarrow T(x_1, y_1) + T(x_2, y_2) = (x_1 + x_2, -y_1 - y_2) = T(x_1 + x_2, y_1 + y_2)$; (b) $\alpha T(x, y) = \alpha(x, -y) = T(\alpha x, \alpha y)$. T is linear. (ii) Let (x, y) be any point in the domain of T, then $T(x, y) = (x, -y) \Rightarrow x = x$, and $y = -y$. Note that if the unique invariant is over the X axis, this axis is the mirror line. T has been reflected in the X axis.

1.2.3. Shear Transformations

A **shear transformation** is a linear transformation that displaces every point in fixed direction, in symbols $(x, y) = (x + \alpha y, y\beta x)$ [Wikipedia, 2017e].

Example 1.18. Be transformation $T: \mathbb{R}^2 \to \mathbb{R}^2$, $(x + 2y, y)$; (i) Is T linear? (ii) Is T a shear?

Solution 1.18. (i) Let (x_1, y_1), $(x_2, y_2) \in \mathbb{R}^2$ be the domain of T, then (a) $T(x_1, y_1) = (x_1 + 2y_1, \ y_1)$, and $T(x_2, y_2) = (x_2 + 2y_2, \ y_2) \Rightarrow T(x_1, y_1) + T(x_2, y_2) = (x_1 + x_2 + 2y_1 + 2y_2, y_1 + y_2) = T(x_1 + x_2, y_1 + y_2) = (x_1 + x_2 + 2y_1 + 2y_2)$; (b) $\alpha T(x, y) = \alpha(x + 2\alpha y, \ y) = T(\alpha x, \alpha y) = (\alpha x + 2\alpha y, y)$. T is linear. (ii) Let (x, y) be any point in the domain of T, then $T(x, y) = (x + \alpha y, y) \Rightarrow x = x + \alpha y$, and $y = y$. T is a shear.

1.3. NONLINEAR TRANSFORMATIONS

Definition 1.10. Suppose A and B are linear spaces over field F. A transformation $T: A \to B$ is said to be **nonlinear** if $\exists\ x, y \in A$, or $\exists\ \alpha \in F$ [Carrell, 2005] Ch. 6, where any of these conditions (1.3, 1.4) are met:

$$T(x + y) \neq T(x) + T(y) \tag{1.3}$$

$$\alpha T(x) \neq T(\alpha x) \tag{1.4}$$

Note 1.3. In Chapter ?? the reader will find this topic more elaborated.

1.3.1. Homothetic Transformations

A **Homothetic transformation** multiplies, from a fixed point, all distances by the same factor.

A **Translation transformation** is a particular type of Homothetic transformation that can be linear or nonlinear. In a translation transformation, all the points in the figure are moved in a straight line in the same direction. This means it preserves the size, shape, and orientation of the original figure in other position.

Example 1.19. Be transformation $T: \mathbb{R}^2 \to \mathbb{R}^2$, $(x + 2, y + 3)$; (i) Is T nonlinear? (ii) Is T a translation?

Solution 1.19. (i) Let (x_1, y_1), $(x_2, y_2) \in \mathbb{R}^2$ be the domain of T, then (a) $T(x_1, y_1) = (x_1 + 2,\ y_1 + 3)$ and $T(x_2, y_2) = (x_2 + 2,\ y_2 + 3) \Rightarrow T(x_1, y_1) + T(x_2, y_2) = (x_1 + x_2 + 4,\ y_1 + y_2 + 6) \neq T(x_1 + x_2, y_1 + y_2) = (x_1 + x_2 + 2, y_1 + y_2 + 3)$; (b) $\alpha T(x, y) = \alpha(x + 2,\ y + 3) \neq T(\alpha x, \alpha y) = (\alpha x + 2, \alpha y + 3)$. T is nonlinear. (ii) Let $T: [0,1] \times [0,1] \subset \mathbb{R}^2 \to \mathbb{R}^2$, $(x + 2, y + 3)$. $T(0,0) = (2,3)$, $T(0,1) = (2,4)$, T(1,1) $=$ (3,4), $T(1,0) = (3,3)$. T is a Homothetic transformation.

Scaling transformation is a particular type of Homothetic transformation, it can be linear or nonlinear. The scaling transformation changes the size of the figure increasing it or decreasing it.

Example 1.20. Be transformation $T: \mathbb{R}^2 \to \mathbb{R}^2$, $(2x, 3y)$; (i) Is T linear? (ii) Is T a scaling?

Solution 1.20. (i) Let (x_1, y_1), $(x_2, y_2) \in \mathbb{R}^2$ be the domain of T, then (a) $T(x_1, y_1) = (2x_1, 2y_1)$, and $T(x_2, y_2) = (2x_2, 2y_2) \Rightarrow T(x_1, y_1) + T(x_2, y_2) = (2x_1 + 2x_2,\ 2y_1 2y_2) = T(x_1 + x_2, y_1 + y_2)$; (b) $\alpha T(x, y) = \alpha(2x,\ 2y) = T(\alpha 2x, \alpha 2y)$. T is linear. (ii) Let $T: [0,1] \times [0,1] \subset \mathbb{R}^2 \to \mathbb{R}^2$, $(2x, 3y)$. $T(0,0) = (0,0)$, $T(0,1) = (0,3)$, T(1,1) $=$ (2,3), $T(1,0) = (2,0)$. T is a scaling.

CHAPTER 2

Linear Transformations

Carlos Polanco[*]

Faculty of Sciences, Universidad Nacional Autónoma de México, México

Abstract: This chapter is focused on the characterization of "linear space" as a mathematical abstraction. It describes the different types of linear spaces, their basic properties, and main components. It also gives a first classification and explains their restrictions and extensions.

Keywords: Coordinate space, Function space, Linear space, Linear transformation, Mapping, Matrix space, Polynomial space, Sequence space, Vector space.

2.1. LINEAR TRANSFORMATIONS

Definition 2.1. Suppose A and B are linear spaces over field F. A transformation $T: A \to B$ is said to be **linear** [Rudin, 1964] if $\forall\ x, y \in A$ and $\forall\ \alpha \in F$ [Carrell, 2005] Ch. 6, such that (2.1, 2.2) are met:

$$T(x + y) = T(x) + T(y) \tag{2.1}$$

$$\alpha T(x) = T(\alpha x) \tag{2.2}$$

Note 2.1. If $T(0) = 0$, $T(x + 0) = T(x) + T(0)$ but $T(x + 0) = T(x) \Rightarrow T(0) = 0$, then $T(0) \neq 0$, this implies transformation T is not linear.

2.1.1. Examples of Linear Transformations

Example 2.1. Show that $T: \mathbb{R} \to \mathbb{R}$, mx is linear.

Solution 2.1. Let $x, y \in \mathbb{R}$ be the domain of T and $m \in F$, then (i) $T(x + y) = m(x + y) = mx + my = T(x) + T(y)$; (ii) $\alpha T(x) = \alpha mx = T(\alpha x)$. T is linear.

Example 2.2. Show that $T: \mathbb{R}^2 \to \mathbb{R}^2$, $(\alpha x, \beta y)$, where α, β are scalars, is linear.

[*]**Corresponding author Carlos Polanco:** Faculty of Sciences, Universidad Nacional Autónoma de México, México City, México; Tel: +01 55 5622 4858; Fax: +01 5556 4859; E-mail: polanco@unam.mx

Solution 2.2. Let (x_1, y_1), $(x_2, y_2) \in \mathbb{R}^2$ be the domain of T, then (i) $T(x_1 + x_2, y_1 + y_2) = [\alpha(x_1 + x_2), \beta(y_1 + y_2)] = (\alpha x_1 + \alpha x_2, \beta y_1 + \beta y_2) = (\alpha x_1, \beta y_1) + (\alpha x_2, \beta y_2) = T(x_1, y_1) + T(x_2, y_2)$; (ii) $\alpha T(x, y) = (\alpha x, \beta y) = T(\alpha x, \beta y)$. T is linear.

Example 2.3. Suppose V is the space of a continuous real valued function. Show that $T: \mathbb{V} \to \mathbb{R}, \int_a^b f(x)\, dx$, is linear [Carrell, 2005] Ch. 6.

Solution 2.3. Let $f, g \in V$, then (i) $T(f + g) = \int_a^b f(x) + g(x)\, dx = \int_a^b f(x)\, dx + \int_a^b g(x)\, dx = T(f) + T(g)$; (ii) $\alpha T(f) = \alpha \int_a^b f(x)\, dx = \int_a^b \alpha f(x)\, dx = T(\alpha f x)$. T is linear.

2.2. LINEAR SPACES

A *linear space* is an important algebraic structure in Linear Algebra where several types of transformations act. In this chapter, we will introduce its definition, the representative linear spaces, as well as several examples and exercises. It is highly recommended that the reader gets familiar with this algebraic structure before reading the following chapters.

Definition 2.2. A *linear space* or "vector space" is an important algebraic structure $L(\Omega, K, +, \times)$ formed by an arbitrary nonempty set Ω, with an internal operation named *vector addition* and an external operation named *scalar multiplication*. The operations between **vectors** from set Ω and **scalars** from a set field K must comply with the following rules $1 - 10$ [Hefferson, 2014] p. 78. That is, a function $K \times \Omega \to \Omega$, denoted by $(\alpha, v) \mapsto \alpha v$, such that for all $\alpha, \beta, 1 \in K$ and all $u, v \in \Omega$ [Rotman, 2002] p. 159.

Rule 1:	$\forall\ v, u \in \Omega$	$\Rightarrow$	$v + u \in \Omega.$
Rule 2:	$\forall\ v, u \in \Omega$	$\Rightarrow$	$v + u = u + v.$
Rule 3:	$\forall\ v, u, w \in \Omega$	$\Rightarrow$	$(u + v) + w = u + (v + w).$
Rule 4:	$\exists!\ 0 \in \Omega, \forall\ v \in \Omega$	$\grave{\text{E}}$	$0 + v = v.$
Rule 5:	$\forall\ v \in \Omega, \quad \exists!\ (-v) \in \Omega$	$\Rightarrow$	$v + (-v) = 0.$
Rule 6:	$\forall\ v \in \Omega, \alpha \in K$	$\Rightarrow$	$\alpha v \in V.$
Rule 7:	$\forall\ v \in \Omega, \alpha, \beta \in K$	$\Rightarrow$	$(\alpha + \beta)v = \alpha v + \beta v.$
Rule 8:	$\forall\ v, w \in \Omega, \alpha \in K$	$\Rightarrow$	$\alpha(v + w) = \alpha v + \alpha w.$
Rule 9:	$\forall\ v \in \Omega, \alpha, \beta \in K$	$\Rightarrow$	$(\alpha\beta)v = \alpha(\beta v).$
Rule 10:	$\forall\ v \in \Omega, \exists!\ 1 \in K$	$\Rightarrow$	$1v = v.$

"The characteristic of arbitrariness", should be widely explained since it plays an important role in the definition of *linear space*. Here are different examples (2.2.1–2.2.3), to highlight this feature with sets that do not comply with it.

2.2.1. Coordinate Space: K^n

Set $\Omega = \{(x_1, x_2, x_3, \cdots, x_n) \in K^n | x_i \in K, i \in \mathbb{N}\}$ is a linear space [Kolmogorov and Fomin, 1970] p. 119.

$$Let \quad v, u \in \Omega; \quad and \quad \alpha \in K$$

$$v + u = (v_1 + u_1, v_2 + u_2, \cdots, v_n + u_n)$$

$$\alpha v = (\alpha v_1, \alpha v_2, \cdots, \alpha v_n)$$

We shall check all of the conditions.

Proof (Rule~1). If $v, u \in \Omega$, where $v = (v_1, v_2, \cdots, v_n)$ and $u = (u_1, u_2, \cdots, u_n)$, then

$$v + u = (v_1, v_2, \cdots, v_n) + (u_1, u_2, \cdots, u_n)$$

$$= (v_1 + u_1, v_2 + u_2, \cdots v_n + u_n) \in \Omega$$

Proof (Rule~2). If $v, u \in \Omega$, where $v = (v_1, v_2, \cdots, v_n)$ and $u = (u_1, u_2, \cdots, u_n)$, then

$$v + u = (v_1, v_2, \cdots, v_n) + (u_1, u_2, \cdots, u_n)$$

$$= (v_1 + u_1, v_2 + u_2, \cdots v_n + u_n)$$

$$= (u_1 + v_1, u_2 + v_2, \cdots u_n + v_n)$$

$$= (u_1, u_2, \cdots, u_n) + (v_1, v_2, \cdots, v_n)$$

$$= u + v$$

Proof (Rule~3). If $v, u, w \in \Omega$, where $v = (v_1, v_2, \cdots, v_n)$, $u = (u_1, u_2, \cdots, u_n)$ and $w = (w_1, w_2, \cdots, w_n)$, then

$$(v + u) + w = ((v_1, v_2, \cdots, v_n) + (u_1, u_2, \cdots, u_n)) +$$
$$(w_1, w_2, \cdots, w_n)$$

$$= (v_1 + u_1 + w_1, v_2 + u_2 + w_2, \cdots v_n + u_n + w_n)$$

$$= (v_1, v_2, \cdots, v_n) + (u_1, u_2, \cdots, u_n) + (w_1, w_2, \cdots, w_n)$$

$$= (v_1, v_2, \cdots, v_n) + ((u_1, u_2, \cdots, u_n) + (w_1, w_2, \cdots, w_n))$$

$$= v + (u + w)$$

Proof (Rule 4). If $v, \theta \in \Omega$, where $v = (v_1, v_2, \cdots, v_n)$, $\theta = (\theta_1, \theta_2, \cdots, \theta_n)$, then

$$v + \theta = v$$

$$(v_1, v_2, \cdots, v_n) + (\theta_1, \theta_2, \cdots, \theta_n) = (v_1, v_2, \cdots, v_n)$$

$$\theta = (0, 0, \cdots, 0)$$

Proof (Rule 4.a). Let $v, \theta \in \Omega$, where $0 = (0, 0, \cdots, 0)$ and $\theta = (\theta_1, \theta_2, \cdots, \theta_n)$, then

$$(v_1, v_2, \cdots, v_n) + (\theta_1, \theta_2, \cdots, \theta_n) = (v_1, v_2, \cdots, v_n)$$

$$(v_1, v_2, \cdots, v_n) + (0, 0, \cdots, 0) = (v_1, v_2, \cdots, v_n)$$

$$(\theta_1, \theta_2, \cdots, \theta_n) = (0, 0, \cdots, 0)$$

Proof (Rule 5). If $v, (-v) \in \Omega$, where $v = (v_1, v_2, \cdots, v_n)$ and $(-v) = -v = (-v_1, -v_2, \cdots, -v_n)$, then

$$v + (-v) = -v + v$$

$$= \theta$$

Proof (Rule 6). If $v \in \Omega$, where $v = (v_1, v_2, \cdots, x_n)$ and $\alpha \in K$, then

$$\alpha v = \alpha(v_1, v_2, \cdots, v_n)$$

$$= (\alpha v_1, \alpha v_2, \cdots, \alpha v_n)$$

$$= ((\alpha v_1), (\alpha v_2), \cdots, (\alpha v_n))$$

$$\alpha v \in \Omega$$

Proof (Rule 7). If $v \in \Omega$, where $v = (v_1, v_2, \cdots, v_n)$ and $\alpha, \beta \in K$, then

$$(\alpha + \beta)v = (\alpha + \beta)(v_1, v_2, \cdots, v_n)$$

$$= ((\alpha + \beta)v_1, (\alpha + \beta)v_2, \cdots, (\alpha + \beta)v_n)$$

$$= (\alpha v_1, \alpha v_2, \cdots, \alpha v_n) + (\beta v_1, \beta v_2, \cdots, \beta v_n)$$

$$= \alpha v + \beta v$$

Proof (Rule 8). If $v, w \in \Omega$, where $v = (v_1, v_2, \cdots, v_n)$, $w = (w_1, w_2, \cdots, w_n)$ and $\alpha \in K$, then

$$\alpha(v + w) = (\alpha v_1 + \alpha w_1, \alpha v_2 + \alpha w_2, \cdots, \alpha v_n + \alpha w_n)$$

$$= (\alpha(v_1 + w_1), \alpha(v_2 + w_2), \cdots, \alpha(v_n + w_n))$$

$$= \alpha v + \alpha w$$

Proof (Rule 9). If $v \in \Omega$, where $v = (v_1, v_2, \cdots, v_n)$ and $\alpha, \beta \in K$, then

$$\alpha\beta(v) = (\alpha\beta v_1, \alpha\beta v_2, \cdots, \alpha\beta v_n)$$

$$= \alpha(\beta v_1 + \beta v_2, \cdots, \beta v_n)$$

$$= \alpha(\beta v)$$

Proof (Rule 10). If $v \in \Omega$, where $v = (v_1, v_2, \cdots, v_n)$ and $1 \in K$, then

$$1v = (1v_1, 1v_2, \cdots, 1v_n)$$

$$= v$$

Note 2.2. There are equivalent notations to express the set of all ordered $n-$ *tuples* as *column vector* [Hefferson, 2014] p. 16, or *row vector* [Malcev, 1963] p. 29.

$$\tilde{v} = \alpha \begin{pmatrix} x_1 \\ x_2 \\ \vdots \\ x_n \end{pmatrix} \qquad \tilde{v} = \alpha(x_1, x_2, \cdots, x_n)$$

2.2.2. Function Space: V^x

The set of all functions $\Omega = \{f(\bar{x}): K^n \to K \,|\, \bar{x} \in K^n\}$ is a linear space.

$$Let \quad f(\bar{x}), \; g(\bar{x}) \in \Omega; \quad and \quad \alpha \in K$$

$$(f + g)(\bar{x}) = f(\bar{x}) + g(\bar{x})$$

$$(\alpha f)(\bar{x}) = \alpha f(\bar{x})$$

We shall check all of the conditions.

Proof (Rule~1). If $f, g \in \Omega$, where $f(\bar{x})$ and $g(\bar{x})$, then

$$(f + g)(\bar{x}) = (f(\bar{x}) + g(\bar{x})) \in \Omega$$

Proof (Rule~2). If $f, g \in \Omega$, where $f(\bar{x})$ and $g(\bar{x})$, then

$$f(\bar{x}) + g(\bar{x}) = f(\bar{x}) + g(\bar{x})$$

$$= g(\bar{x}) + f(\bar{x})$$

Proof (Rule~3). If $f, g, w \in \Omega$, where $f(\bar{x}), g = (\;x)$ and $w(\bar{x})$, then

$$(f(\bar{x}) + g(\bar{x})) + w(\bar{x}) = (f(\bar{x}) + g(\bar{x})) + w(\bar{x})$$

$$= f(\bar{x}) + g(\bar{x}) + w(\bar{x})$$

$$= f(\bar{x}) + (g(\bar{x}) + w(\bar{x}))$$

Proof (Rule 4). If $f, \; 0, \; \theta \in \Omega$, where $f(\bar{x}), \; \theta(\bar{x})$ and $0(\bar{x})$, then

$$f(\bar{x}) + \theta(\bar{x}) = f(\bar{x})$$

$$\theta(\bar{x}) = 0(\bar{x})$$

Proof (Rule 4.a). Let $f, \; 0, \; \theta \in \Omega$, where $f(\bar{x}), \; 0(\bar{x})$ and $\theta(\bar{x})$, then

$$f(\bar{x}) + \theta(\bar{x}) = f(\bar{x})$$

$$f(\bar{x}) + 0(x) = f(\bar{x})$$

$$\theta(\bar{x}) = 0(x)$$

Proof (Rule 5). If $f, (-f)(\bar{x}) \in \Omega$, where $f(\bar{x})$ and $(-)f(\bar{x}) = -f(\bar{x})$, then

$$-f(\bar{x}) + f(\bar{x}) = (-f)(\bar{x}) + f(\bar{x})$$

$$= -f(\bar{x}) + f(\bar{x})$$

$$= \theta$$

Proof (Rule 6). If $f \in \Omega$, where $f(\bar{x})$ and $\alpha \in K$, then

$$\alpha f(\bar{x}) = \alpha f(\bar{x})$$

$$\alpha f(\bar{x}) \in \Omega$$

Proof (Rule 7). If $f \in \Omega$, where $f(\bar{x})$ and $\alpha, \beta \in K$, then

$$(\alpha + \beta)f(\bar{x}) = (\alpha + \beta)f(\bar{x})$$

$$= \alpha f(\bar{x}) + \beta f(\bar{x})$$

Proof (Rule 8). If $f, g \in \Omega$, where $f(\bar{x})$, $g(\bar{x})$ and $\alpha \in K$, then

$$\alpha(f(\bar{x}) + g(\bar{x})) = \alpha f(\bar{x}) + \beta g(\bar{x})$$

Proof (Rule 9). If $f \in \Omega$, where $f(\bar{x})$ and $\alpha, \beta \in K$, then

$$\alpha\beta f(\bar{x}) = (\alpha\beta)f(\bar{x})$$

$$= \alpha(\beta f(\bar{x}))$$

Proof (Rule 10). If $f \in \Omega$, where $f(\bar{x})$ and $1 \in K$, then

$$1f(\bar{x}) = 1f(\bar{x})$$

$$= f(\bar{x})$$

2.2.3. Polynomial Space: P_n

The set of polynomials $\Omega = \{p(x) : K \to K \mid p(x) = a_0 x^0 + a_1 x^1 +, \cdots, + a_n x^n,\ a_i \in K,\ i \in \mathbb{N}\}$ with degree less than or equal to "n", is a linear space.

$$Let \quad p, q \in \Omega; \quad and \quad \alpha, a_i, b_i \in K$$

$$p(x) + q(x) = (a_0 + b_0)x^0 + (a_1 + b_1)x^1 +, \cdots, +(a_n + b_n)x^n$$

$$\alpha p(x) = \alpha a_0 x^0 + \alpha a_1 x^1 +, \cdots, \alpha a_n x^n$$

We shall check all of the conditions.

Proof (Rule~1). If $p, q \in \Omega$, where $p(x) = a_0 x^0 + a_1 x^1 +, \cdots, +a_n x^n$ and $q(x) = b_0 x^0 + b_1 x^1 +, \cdots, +b_n x^n$, then

$$(p + q)(x) = (a_0 + b_0)x^0, +(a_1 + b_1)x^1 +, \cdots, +(a_n + b_n)x^n \in \Omega$$

Proof (Rule~2). If $p, q \in \Omega$, where $p(x) = a_0 x^0 + a_1 x^1 +, \cdots, +a_n x^n$ and $q(x) = b_0 x^0 + b_1 x^1 +, \cdots, +b_n x^n$, then

$$p(x) + q(x) = (a_0 + b_0)x^0 + (a_1 + b_1)x^1 +, \cdots, +(a_n + b_n)x^n$$

$$= (b_0 + a_0)x^0 + (b_1 + a_1)x^1 +, \cdots, +(b_n + a_n)x^n$$

$$= q(x) + p(x)$$

Proof (Rule~3). If $p, q, r \in \Omega$, where $p(x) = a_0 x^0 + a_1 x^1 +, \cdots, +a_n x^n$, $q(x) = b_0 x^0 + b_1 x^1 +, \cdots, +b_n x^n$ and $r(x) = c_0 x^0 + c_1 x^1 +, \cdots, +c_n x^n$, then

$$((p(x) + q(x)) + r(x) = (a_0 + b_0 + c_0)x^0 + (a_1 + b_1 + c_1)x^1 +, \cdots, +(a_n + b_n + c_n)x^n$$

$$= [a_0 + (b_0 + c_0)]x^0 + [a_1 + (b_1 + c_1)]x^1 +, \cdots, +[a_n + (b_n + c_n)]x^n$$

$$= p(x) + (q(x) + r(x))$$

Proof (Rule 4). If $p, \theta \in \Omega$, where $p(x) = a_0 x^0 + a_1 x^1 +, \cdots, +a_n x^n$, $\theta(x) = \theta_0 x^0 + \theta_1 x^1 +, \cdots, +\theta_n x^n$ and $0(x) = 0x^0 + 0x^1 +, \cdots, +0x^n$, then

$$p(x) + \theta(x) = p(x)$$

$$\theta(x) = 0(x)$$

Proof (Rule 4.a). Let $p, \theta \in \Omega$, where $p(x) = a_0 x^0 + a_1 x^1 +, \cdots, + a_n x^n$, $\theta(x) = \theta_0 x^0 + \theta_1 x^1 +, \cdots, + \theta_n x^n$ and $0(x) = 0x^0 + 0x^1 +, \cdots, + 0x^n$, then

$$p(x) + \theta(x) = p(x)$$

$$p(x) + 0(x) =$$

$$\theta(x) = 0(x)$$

Proof (Rule 5). If $p, (-p) \in \Omega$, where $p(x) = a_0 x^0 + a_1 x^1 +, \cdots, + a_n x^n$, $(-p)(x) = -p(x) = -a_0 x^0 - a_1 x^1, \cdots, -a_n x^n$ and $\theta(x) = \theta_0 x^0 + \theta_1 x^1 +, \cdots, + \theta_n x^n$, .then

$$(-p) + p(x) = -p(x) + p(x)$$

$$= (-a_0 + a_0)x^0 + (-a_1 + a_1)x^1 +, \cdots, + (-a_n + a_n)x^n$$

$$= 0x^0 + 0x^1 +, \cdots, + 0x^n$$

$$= \theta$$

Proof (Rule 6). If $p \in \Omega$, where $p(x) = a_0 x^0 + a_1 x^1 +, \cdots, + a_n x^n$ and $\alpha \in K$, then

$$\alpha p(x) = \alpha p(x)$$

$$= \alpha a_0 x^0 + \alpha a_1 x^1 +, \cdots, + \alpha a_n x^n$$

$$\alpha p(x) \in \Omega$$

Proof (Rule 7). If $p \in \Omega$, where $p(x) = a_0 x^0 + a_1 x^1 +, \cdots, + a_n x^n$ and $\alpha, \beta \in K$, then

$$(\alpha + \beta)p(x) = (\alpha + \beta)p(x)$$

$$= (\alpha + \beta)a_0 x^0, (\alpha + \beta)a_1 x^1, \cdots, (\alpha + \beta)a_n x^n$$

$$= \alpha p(x) + \beta p(x)$$

Proof (Rule 8). If $p, q \in \Omega$, where $p(x) = a_0 x^0 + a_1 x^1 +, \cdots, + a_n x^n$, $q(x) = b_0 x^0 + b_1 x^1 +, \cdots, + b_n x^n$ and $\alpha \in K$, then

$$\alpha(p(x) + q(x)) = \alpha(a_0 + b_0)x^0 + \alpha(a_1 + b_1)x^1+, \cdots, +\alpha(a_n + b_n)x^n$$

$$= \alpha p(x) + \alpha q(x)$$

Proof (Rule 9). If $p \in \Omega$, where $p(x) = a_0 x^0 + a_1 x^1+, \cdots, +a_n x^n$ and $\alpha, \beta \in K$, then

$$\alpha\beta(p(x)) = \alpha\beta a_0 x^0 + \alpha\beta a_1 x^1 + \cdots, +\alpha\beta a_n x^n$$

$$= \alpha(\beta(p(x)))$$

Proof (Rule 10). If $p \in \Omega$, where $p(x) = a_0 x^0 + a_1 x^1+, \cdots, +a_n x^n$ and $1 \in K$, then

$$1p(x) = 1a_0 x^0 + 1a_1 x^1+, \cdots, +1a_n x^n$$

$$= p(x)$$

2.2.4. Matrix Space: $M_{m,n}$

The set of all matrices $\Omega = \{M_{m,n}: K^{m \times n} \to K^{m \times n} | x_{i,j} \in K, \ i,j \in \mathbb{N}\}$, is a linear space.

Let $\quad A, B \in \Omega; \quad and \quad \alpha, \ A(i,j), \ B(i,j) \in K, i \in [1, \cdots, m], j \in [1, \cdots, n]$

$$A + B = \begin{pmatrix} a_{11} + b_{11} & a_{12} + b_{12} & \cdots & a_{1n} + b_{1n} \\ a_{21} + b_{21} & a_{22} + b_{22} & \cdots & a_{2n} + b_{2n} \\ \vdots & \vdots & \ddots & \vdots \\ a_{m1} + b_{m1} & a_{m2} + b_{m2} & \cdots & a_{mn} + b_{mn} \end{pmatrix}$$

$$\alpha A = \begin{pmatrix} \alpha a_{11} & \alpha a_{12} & \cdots & \alpha a_{1n} \\ \alpha a_{21} & \alpha a_{22} & \cdots & \alpha a_{2n} \\ \vdots & \vdots & \ddots & \vdots \\ \alpha a_{m1} & \alpha a_{m2} & \cdots & \alpha a_{mn} \end{pmatrix}$$

We shall check all of the conditions.

Proof (Rule~1). If $A, B \in \Omega$, where $A(a_{i,j}), B(b_{i,j})$, then

$$A + B = \begin{pmatrix} a_{11} + b_{11} & a_{12} + b_{12} & \cdots & a_{1n} + b_{1n} \\ a_{21} + b_{21} & a_{22} + b_{22} & \cdots & a_{2n} + b_{2n} \\ \vdots & \vdots & \ddots & \vdots \\ a_{m1} + b_{m1} & a_{m2} + b_{m2} & \cdots & a_{mn} + b_{mn} \end{pmatrix} \in \Omega$$

Proof (Rule~2). If $A, B \in \Omega$, where $A(a_{i,j}), B(b_{i,j})$, then

$$A + B = \begin{pmatrix} a_{11} + b_{11} & a_{12} + b_{12} & \cdots & a_{1n} + b_{1n} \\ a_{21} + b_{21} & a_{22} + b_{22} & \cdots & a_{2n} + b_{2n} \\ \vdots & \vdots & \ddots & \vdots \\ a_{m1} + b_{m1} & a_{m2} + b_{m2} & \cdots & a_{mn} + b_{mn} \end{pmatrix}$$

$$= \begin{pmatrix} b_{11} + a_{11} & b_{12} + a_{12} & \cdots & b_{1n} + a_{1n} \\ b_{21} + a_{21} & b_{22} + a_{22} & \cdots & b_{2n} + a_{2n} \\ \vdots & \vdots & \ddots & \vdots \\ b_{m1} + a_{m1} & b_{m2} + a_{m2} & \cdots & b_{mn} + a_{mn} \end{pmatrix}$$

$$= B + A$$

Proof (Rule~3). If $A, B, C \in \Omega$, where $A(a_{i,j}), B(b_{i,j})$ and $C(c_{i,j})$, then

$(A + B) + C$

$$= \begin{pmatrix} (a_{11} + b_{11}) + c_{11} & (a_{12} + b_{12}) + c_{11} & \cdots & (a_{1n} + b_{1n}) + c_{1n} \\ (a_{21} + b_{21}) + c_{11} & (a_{22} + b_{22}) + c_{22} & \cdots & (a_{2n} + b_{2n}) + c_{1n} \\ \vdots & \vdots & \ddots & \vdots \\ (a_{m1} + b_{m1}) + c_{m1} & (a_{m2} + b_{m2}) + c_{m2} & \cdots & (a_{mn} + b_{mn}) + c_{mn} \end{pmatrix}$$

$$= \begin{pmatrix} a_{11} + (b_{11} + c_{11}) & a_{12} + (b_{12} + c_{12}) & \cdots & a_{1n} + (b_{1n} + c_{1n}) \\ a_{21} + (b_{21} + c_{21}) & a_{22} + (b_{22} + c_{22}) & \cdots & a_{2n} + (b_{2n} + c_{2n}) \\ \vdots & \vdots & \ddots & \vdots \\ a_{m1} + (b_{m1} + c_{m1}) & a_{m2} + (b_{m2} + c_{m2}) & \cdots & a_{mn} + (b_{mn} + c_{mn}) \end{pmatrix}$$

$$= A + (B + C)$$

Proof (Rule~4). If $A, \theta \in \Omega$, where $A(a_{i,j}), \theta(\theta_{i,j})$, then

$$
A + \theta = \begin{pmatrix}
a_{11} + \theta_{11} & a_{12} + \theta_{12} & \cdots & a_{1n} + \theta_{1n} \\
a_{21} + \theta_{21} & a_{22} + \theta_{22} & \cdots & a_{2n} + \theta_{2n} \\
\vdots & \vdots & \ddots & \vdots \\
a_{m1} + \theta_{m1} & a_{m2} + \theta_{m2} & \cdots & a_{mn} + \theta_{mn}
\end{pmatrix}
$$

$$
= \begin{pmatrix}
a_{11} & a_{12} & \cdots & a_{1n} \\
a_{21} & a_{22} & \cdots & a_{2n} \\
\vdots & \vdots & \ddots & \vdots \\
a_{m1} & a_{m2} & \cdots & a_{mn}
\end{pmatrix}
$$

$$\theta = 0$$

Proof (Rule~4.a). Let $A, \theta \in \Omega$, where $A(a_{i,j}), \theta(\theta_{i,j})$, then

$$
A + \theta = \begin{pmatrix}
a_{11} + \theta_{11} & a_{12} + \theta_{12} & \cdots & a_{1n} + \theta_{1n} \\
a_{21} + \theta_{21} & a_{22} + \theta_{22} & \cdots & a_{2n} + \theta_{2n} \\
\vdots & \vdots & \ddots & \vdots \\
a_{m1} + \theta_{m1} & a_{m2} + \theta_{m2} & \cdots & a_{mn} + \theta_{mn}
\end{pmatrix}
$$

$$
= \begin{pmatrix}
a_{11} & a_{12} & \cdots & a_{1n} \\
a_{21} & a_{22} & \cdots & a_{2n} \\
\vdots & \vdots & \ddots & \vdots \\
a_{m1} & a_{m2} & \cdots & a_{mn}
\end{pmatrix}
$$

$$
A + 0 = \begin{pmatrix}
a_{11} + 0_{11} & a_{12} + 0_{12} & \cdots & a_{1n} + 0_{1n} \\
a_{21} + 0_{21} & a_{22} + 0_{22} & \cdots & a_{2n} + 0_{2n} \\
\vdots & \vdots & \ddots & \vdots \\
a_{m1} + 0_{m1} & a_{m2} + 0_{m2} & \cdots & a_{mn} + 0_{mn}
\end{pmatrix}
$$

$$
= \begin{pmatrix}
a_{11} & a_{12} & \cdots & a_{1n} \\
a_{21} & a_{22} & \cdots & a_{2n} \\
\vdots & \vdots & \ddots & \vdots \\
a_{m1} & a_{m2} & \cdots & a_{mn}
\end{pmatrix}
$$

$$\theta = 0$$

Proof (Rule~5). If $A, -A \in \Omega$, where $A(a_{i,j})$ and $-A(a_{i,j}) = A(-a_{i,j})$, then

$$(-A) + A = \begin{pmatrix} -a_{11} + a_{11} & -a_{12} + a_{12} & \cdots & -a_{1n} + a_{1n} \\ -a_{21} + a_{21} & -a_{22} + a_{22} & \cdots & -a_{2n} + a_{2n} \\ \vdots & \vdots & \ddots & \vdots \\ -a_{m1} + a_{m1} & -a_{m2} + a_{m2} & \cdots & -a_{mn} + a_{mn} \end{pmatrix}$$

$$= \begin{pmatrix} 0_{11} & 0_{12} & \cdots & 0_{1n} \\ 0_{21} & 0_{22} & \cdots & 0_{2n} \\ \vdots & \vdots & \ddots & \vdots \\ 0_{m1} & 0_{m2} & \cdots & 0_{mn} \end{pmatrix}$$

$$= \theta$$

Proof (Rule~6). If $A \in \Omega$, where $A(a_{i,j})$ and $\alpha \in K$, then

$$\alpha A = \begin{pmatrix} \alpha a_{11} & \alpha a_{12} & \cdots & \alpha a_{1n} \\ \alpha a_{21} & \alpha a_{22} & \cdots & \alpha a_{2n} \\ \vdots & \vdots & \ddots & \vdots \\ \alpha a_{m1} & \alpha a_{m2} & \cdots & \alpha a_{mn} \end{pmatrix}$$

$$\alpha A \in \Omega$$

Proof (Rule~7). If $A \in \Omega$, where $A(a_{i,j})$ and $\alpha, \beta \in K$, then

$$(\alpha + \beta)A = \begin{pmatrix} (\alpha + \beta)a_{11} & (\alpha + \beta)a_{12} & \cdots & (\alpha + \beta)a_{1n} \\ (\alpha + \beta)a_{21} & (\alpha + \beta)a_{22} & \cdots & (\alpha + \beta)a_{2n} \\ \vdots & \vdots & \ddots & \vdots \\ (\alpha + \beta)a_{m1} & (\alpha + \beta)a_{m2} & \cdots & (\alpha + \beta)a_{mn} \end{pmatrix}$$

$$= \begin{pmatrix} \alpha a_{11} + \beta a_{11} & \alpha a_{12} + \beta a_{12} & \cdots & \alpha a_{1n} + \beta a_{1n} \\ \alpha a_{21} + \beta a_{21} & \alpha a_{22} + \beta a_{22} & \cdots & \alpha a_{2n} + \beta a_{2n} \\ \vdots & \vdots & \ddots & \vdots \\ \alpha a_{m1} + \beta a_{m1} & \alpha a_{m2} + \beta a_{m2} & \cdots & \alpha a_{mn} + \beta a_{mn} \end{pmatrix}$$

$$= \alpha A + \beta A$$

Proof (Rule~8). If $A, B \in \Omega$, where $A(a_{i,j})$ and $\alpha \in K$, then

$$\alpha(A + B) = \begin{pmatrix} \alpha(a_{11} + b_{11}) & \alpha(a_{12} + b_{12}) & \cdots & \alpha(a_{1n} + b_{1n}) \\ \alpha(a_{21} + b_{21}) & \alpha(a_{22} + b_{22}) & \cdots & \alpha(a_{2n} + b_{2n}) \\ \vdots & \vdots & \ddots & \vdots \\ \alpha(a_{m1} + b_{m1}) & \alpha(a_{m2} + b_{m2}) & \cdots & \alpha(a_{mn} + b_{mn}) \end{pmatrix}$$

$$= \begin{pmatrix} \alpha a_{11} + \alpha b_{11} & \alpha a_{12} + \alpha b_{12} & \cdots & \alpha a_{1n} + \alpha b_{1n} \\ \alpha a_{21} + \alpha b_{21} & \alpha a_{22} + \alpha b_{22} & \cdots & \alpha a_{2n} + \alpha b_{2n} \\ \vdots & \vdots & \ddots & \vdots \\ \alpha a_{m1} + \alpha b_{m1} & \alpha a_{m2} + \alpha b_{m2} & \cdots & \alpha a_{mn} + \alpha b_{mn} \end{pmatrix}$$

$$= \alpha A + \alpha B$$

Proof (Rule~9). If $A \in \Omega$, where $A(a_{i,j})$ and $\alpha, \beta \in K$, then

$$(\alpha\beta)A = \begin{pmatrix} \alpha\beta a_{11} & \alpha\beta a_{12} & \cdots & \alpha\beta a_{1n} \\ \alpha\beta a_{21} & \alpha\beta a_{22} & \cdots & \alpha\beta a_{2n} \\ \vdots & \vdots & \ddots & \vdots \\ \alpha\beta a_{m1} & \alpha\beta a_{m2} & \cdots & \alpha\beta a_{mn} \end{pmatrix}$$

$$= \begin{pmatrix} \alpha(\beta a_{11}) & \alpha(\beta a_{12}) & \cdots & \alpha(\beta a_{1n}) \\ \alpha(\beta a_{21}) & \alpha(\beta a_{22}) & \cdots & \alpha(\beta a_{2n}) \\ \vdots & \vdots & \ddots & \vdots \\ \alpha(\beta a_{m1}) & \alpha(\beta a_{m2}) & \cdots & \alpha(\beta a_{mn}) \end{pmatrix}$$

$$= \alpha(\beta A)$$

Proof (Rule~10). If $A, \in \Omega$, where $A(a_{i,j})$ and $1 \in K$, then

$$(1)A = \begin{pmatrix} 1a_{11} & 1a_{12} & \cdots & 1a_{1n} \\ 1a_{21} & 1a_{22} & \cdots & 1a_{2n} \\ \vdots & \vdots & \ddots & \vdots \\ 1a_{m1} & 1a_{m2} & \cdots & 1a_{mn} \end{pmatrix} = A$$

2.2.5. Sequence Space: $(a_n)_{n \in \mathbb{N}}$

The set of functions $\Omega = \{f(i): \mathbb{N} \to K|$, where element i is a_i, $i \in \mathbb{N}\}$ is a linear space.

It is important to note that there are different expressions for a succession, here a sequence is to be understood as the term a_i, which corresponds to the position i generated by function $f(i)$, *i.e.* $f(i) = (a_i)_{i \in \mathbb{N}} = a_i$.

$$Let \quad f, g \in \Omega; \quad and \quad \alpha \in K$$

$$f(i) + g(i) = a_i + b_i$$

$$\alpha f(i) = \alpha a_i$$

We shall check all of the conditions.

Proof (Rule~1). If $f, g \in \Omega$, where $f(i) = a_i$ and $g(i) = b_i$, then

$$(f + g)(i) = f(i) + g(i) = a_i + b_i \in \Omega$$

Proof (Rule~2). If $f, g \in \Omega$, where $f(i) = a_i$ and $g(i) = b_i$, then

$$f(i) + g(i) = a_i + b_i$$

$$= b_i + a_i$$

$$= g(i) + f(i)$$

Proof (Rule~3). If $f, g, w \in \Omega$, where $f(i) = a_i$, $g = (i) = b_i$ and $w(i) = c_i$, then

$$(f(i) + g(i)) + w(i) = (f(i) + g(i)) + w(i)$$

$$= f(i) + g(i) + w(i)$$

$$= a_i + (b_i + c_i)$$

$$= f(i) + (g(i) + w(i))$$

Proof (Rule 4). If f, $\theta \in \Omega$, where $f(i) = a_i$ and $\theta(i) = \theta_i$, then

$$f(i) + \theta(i) = f(i)$$

$$a_i + \theta_i = a_i$$

$$\theta(i) = 0$$

Proof (Rule 4.a). Let $f, \theta \in \Omega$, where $f(i) = a_i$ and $\theta(i) = \theta_i$, then

$$f(i) + \theta(i) = f(i)$$

$$f(i) + 0 = f(i)$$

$$a_i + 0 = a_i$$

$$\theta(i) = 0$$

Proof (Rule 5). If $f, (-f)(i) \in \Omega$, where $f(i) = a_i$ and $(-)f(i) = -f(i) = -a_i$, then

$$-f(i) + f(i) = -a_i + a_i$$

$$= \theta$$

Proof (Rule 6). If $f \in \Omega$,where $f(i) = a_i$ and $\alpha \in K$, then

$$\alpha f(i) = \alpha a_i$$

$$\alpha f(i) \in \Omega$$

Proof (Rule 7). If $f \in \Omega$, where $f(i) = a_i$ and $\alpha, \beta \in K$, then

$$(\alpha + \beta)f(i) = (\alpha + \beta)a_i$$

$$= \alpha a_i + \beta a_i$$

$$= \alpha f(i) + \beta f(i)$$

Proof (Rule 8). If $f, g \in \Omega$, where $f(i) = a_i$, $g(i)i = b_i$ and $\alpha \in K$, then

$$\alpha(f(i) + g(i)) = \alpha(a_1 + b_i)$$

$$= \alpha a_i + \alpha b_i$$

$$= \alpha f(i) + \alpha g(i)$$

Proof (Rule 9). If $f \in \Omega$, where $f(i) = a_i$ and $\alpha, \beta \in K$, then

$$\alpha \beta f(i) = (\alpha \beta) f(i)$$

$$= (\alpha \beta) a_i$$

$$= \alpha (\beta a_i)$$

$$= \alpha (\beta f(i))$$

Proof (Rule 10). If $f \in \Omega$, where $f(i) = a_i$ and $1 \in K$, then

$$1 f(i) = 1 f(i)$$

$$= 1 a_i$$

$$= f(i)$$

2.2.6. Examples of Linear Spaces

Example 2.4. Is the coordinate space defined by set $\Omega = \{(x_1, x_2) \in K^2 | x_i \in K, i \in \mathbb{N}\}$ a linear space? [Rojo, 1973] p. 4.

$$Let \quad v, u \in \Omega; \quad and \quad \alpha \in K$$

$$v + u = (v_1 + u_1, v_2 + u_2)$$

$$\alpha v = (v_1, v_1)$$

Solution 2.4. No, it is not. Rule 7: $(\alpha + \beta)v = \alpha v + \beta v$, is not met. Left side:

$$(\alpha + \beta)v = (\alpha + \beta)(v_1, v_2)$$

$$= (v_1, v_1)$$

Right side:

$$\alpha v + \beta v = \alpha(v_1, v_2) + \beta(v_1, v_2)$$

$$= (v_1 + v_1) + (v_1, v_1)$$

$$= (2v_1, 2v_1)$$

Example 2.5. Is the coordinate space defined by the set $\Omega = \{(x_1, x_2) \in K^2 | x_i \in K, i \in \mathbb{N}\}$ a linear space? [Friedberg *et al.*, 1982].

$$Let \quad v, u \in \Omega; \quad and \quad \alpha \in K$$

$$v + u = (v_1 + u_1, v_2 u_2)$$

$$\alpha v = (\alpha v_1, v_2)$$

Solution 2.5. No, it is not. Rule 5: $v + (-v) = 0$, is not met.

$$v + (-v) = 0$$

$$= (v_1, v_2) + (-v_1, -v_2$$

$$= (v_1 - v_1, -v_2 v_2)$$

$$= (0, -v_2^2)$$

Example 2.6. Is the set of polynomials of degree "n", under the standard operations of addition and multiplication by scalars, a linear space? [Malcev, 1963] p. 34.

$$Let \quad p, q \in \Omega; \quad and \quad \alpha, a_i, b_i \in K$$

$$p(x) + q(x) = (a_0 + b_0)x^0 + (a_1 + b_1)x^1 +, \cdots, +(a_n + b_n)x^n$$

$$\alpha p(x) = \alpha a_0 x^0 + \alpha a_1 x^1 + \cdots, + \alpha a_n x^n$$

Solution 2.6. No, it is not. Rule 1: $v + u \in \Omega$, is not met.

Since $p, q \in \Omega$, where $p(x) = a_0 x^0 + a_1 x^1 +, \cdots, +a_n x^n$ and $q(x) = b_0 x^0 + b_1 x^1 +, \cdots, +b_m x^m$, where $m < n$. Then,

$$(p + q)(x) = (a_0 + b_0)x^0, (a_1 + b_1)x^1 +, \cdots, +(a_m + b_m)x^m \cdots, +(a_n +?)x^n$$

$$(p + q)(x) \notin \Omega$$

Example 2.7. Is the singleton set Ω a linear space, under the standard operations of addition and multiplication on matrix space $M_{m,n}$? [Hefferson, 2014] p. 82.

$$\Omega = \{ \begin{pmatrix} 0 \\ 0 \\ \vdots \\ 0 \end{pmatrix} \}$$

Solution 2.7. Yes, it is. All rules are met. A linear space must have at least one element, its zero vector. Thus, a one-element linear space is the smallest possible.

Definition 2.3. A one-element linear space is a *trivial space*.

Example 2.8. Is the set $\Omega = \{f(x) \in \mathbb{R} | f(x): \mathbb{R} \to \mathbb{R} | \frac{d^2 f}{dx^2} + f = 0\}$ a linear space? [Hefferson, 2014] p. 85.

$$Let \quad f, g \in \Omega; \quad and \quad \alpha \in \mathbb{R}$$

$$(f + g)(x) = f(x) + g(x)$$

$$(\alpha f)(x) = \alpha f(x)$$

Solution 2.8. Yes, it is. All rules are met.

Proof. Rule 1. $\frac{d^2(f+g)}{dx^2} + (f + g) \Rightarrow \frac{d^2 f}{dx^2} + f + \frac{d^2 g}{dx^2} + g$

Rule 2. $\frac{d^2(f+g)}{dx^2} + (f + g) \Rightarrow \frac{d^2 f}{dx^2} + f + \frac{d^2 g}{dx^2} + g \Rightarrow \frac{d^2 g}{dx^2} + g + \frac{d^2 f}{dx^2} + f \Rightarrow$ $\frac{d^2(g+f)}{dx^2} + (g + f)$

Rule 3. $\frac{d^2((f+g)+h)}{dx^2} + ((f + g) + h) \Rightarrow (\frac{d^2 f}{dx^2} + f + \frac{d^2 g}{dx^2} + g) + \frac{d^2 h}{dx^2} + h \Rightarrow \frac{d^2 f}{dx^2} +$ $f + (\frac{d^2 g}{dx^2} + g + \frac{d^2 h}{dx^2} + h) \Rightarrow \frac{d^2(f+(g+h))}{dx^2} + (f + (g + h))$

Rule 4. $\frac{d^2(f+\theta)}{dx^2} + (f + \theta) \Rightarrow \frac{d^2 f}{dx^2} + f \frac{d^2 \theta}{dx^2} + \theta \Rightarrow \frac{d^2 0}{dx^2} + 0$

Rule 4.a. $\frac{d^2(f+\theta)}{dx^2} + (f + \theta) \Rightarrow \frac{d^2f}{dx^2} + f \, \frac{d^2(f+0)}{dx^2} + (f + 0) \Rightarrow \frac{d^2f}{dx^2} + f \, \frac{d^2\theta}{dx^2} + \theta \Rightarrow$ $\frac{d^2 0}{dx^2} + 0$

Rule 5. $\frac{d^2(-f+f)}{dx^2} + (-f + f) \Rightarrow \frac{d^2-f}{dx^2} - f + \frac{d^2f}{dx^2} + f \Rightarrow \frac{d^2\theta}{dx^2} + \theta$

Rule 6. $\alpha(\frac{d^2f}{dx^2} + f) \Rightarrow \frac{d^2\alpha f}{dx^2} + \alpha f$

Rule 7. $(\alpha + \beta)(\frac{d^2f}{dx^2} + f) \Rightarrow \alpha(\frac{d^2f}{dx^2} + f) + \beta(\frac{d^2f}{dx^2} + f)$

Rule 8. $\alpha \frac{d^2(f+g)}{dx^2} + (f + g) \Rightarrow \alpha(\frac{d^2f}{dx^2} + f) + \alpha(\frac{d^2g}{dx^2} + g)$

Rule 9. $\alpha\beta(\frac{d^2f}{dx^2} + f) \Rightarrow \alpha(\beta(\frac{d^2f}{dx^2} + f))$

Rule 10. $1(\frac{d^2f}{dx^2} + f) \Rightarrow \frac{d^2f}{dx^2} + f$

2.3. LINEAR INDEPENDENCE

Definition 2.4. A finite system of vectors $B = v_1, v_2, \cdots, v_n$ from a linear space V is called **linearly independent** if $\alpha_1 v_1 + \alpha_2 v_2, \cdots \alpha_n v_n = 0$, where $\alpha_i \in$ field F implies that $\alpha_i = 0$ [Malcev, 1963] p. 35. **Basis** B of *linear space* V over field F is a *linearly independent* subset of V that **spans** V. The spanning property means that for every $x \in V$ it is possible to choose $\alpha_i F$ such that $x = \alpha_1 v_1 + \alpha_2 v_2, \cdots \alpha_n v_n$ [Wikipedia, 2017b].

Note 2.3. Bases are not unique. While there exists a unique way to express a vector in terms of any particular basis, bases themselves are far from unique. For example, both sets $\{(1,0), (0,1)\}$ and $\{(1,1), (1,-1)\}$ are bases for $\mathbb{R}^2$ [Cherney *et al.*, 2013]. In all cases, it is necessary to proof the linear and spanning properties.

2.3.1. Examples of Linear Independence

Example 2.9. Two vectors u and v are linearly independent if the only numbers x and y satisfying $xu + yv = 0$ are $x = y = 0$ [Thompson, 1996], Section Testing for Linear Dependence of Vectors).

Proof. Be

$$u = \begin{pmatrix} a \\ b \end{pmatrix} \quad \text{and,} \quad v = \begin{pmatrix} c \\ d \end{pmatrix} \quad \Rightarrow \quad 0 = x \begin{pmatrix} a \\ b \end{pmatrix} + y \begin{pmatrix} c \\ d \end{pmatrix} = \begin{pmatrix} a & c \\ b & d \end{pmatrix} \begin{pmatrix} x \\ y \end{pmatrix}$$

"If u and v are linearly independent, then the only solution to this system of equations is the trivial solution, $x = y = 0$. For homogeneous systems, this happens precisely when the determinant is non-zero. We have now found a test for determining whether a given set of vectors is linearly independent: A set of n vectors of length n is linearly independent if the matrix with these vectors has as columns a non-zero determinant. The set will be dependent if the determinant is zero" [Thompson, 1996], Section Testing for Linear Dependence of Vectors).

2.4. LINEAR SPAN

Given a vector space V over a field F, the **span** of a set A of vectors will be defined by $A = \{\sum_{i=1}^{m} \alpha_i v_i \mid m \in \mathbb{N}, v_i \in V, \alpha_i \in \text{field } F\}$. Particularly, if A is a finite subset of linear space V, then the span of A will be the set of all linear combinations of the elements of A [Wikipedia, 2017c].

2.4.1. Examples of Linear Span

Example 2.10. Is the set $\{(-1,0,0), (0,1,0), (0,0,1)\}$ a spanning set of real vector space $\mathbb{R}^3$ [Wikipedia, 2017c]?

Solution 2.9. Yes it is. The real linear space $\mathbb{R}^3$ has $\{(-1,0,0), (0,1,0), (0,0,1)\}$ as a spanning set. This particular spanning set is also a *basis*.

Note 2.4. If $(-1,0,0)$ were replaced by $(1,0,0)$, it would also form the canonical basis of $\mathbb{R}^3$.

Example 2.11. Is the spanning set given by $\{(1,2,3), (0,1,2), (1,2,3), (1,1,1)\}$, basis of linear space $\mathbb{R}^3$? [Wikipedia, 2017c].

Solution 2.10. No, it is not. The spanning set $\{(1,2,3), (0,1,2), (1,2,3), (1,1,1)\}$, is linearly dependent.

2.5. INVARIANT SUBSPACES

Suppose $T: V \to V$ is a linear transformation and S is a subspace of V. Furthermore, suppose $T(s) \subseteq S$ for every $s \in S$. Then S is an invariant subspace of V relative to T ?, [Beezer, 2017; Cullen, 2003].

Definition 2.5. S subset of V is a subspace of V if: (a) there is a zero element in S, (b) s_1 and $s_2 \in S$, then $s_1 + s_2 \in S$, and (c) $\alpha \in \mathbb{R}$, $\alpha s \in S$.

2.5.1. Examples of Invariant Subspaces

Example 2.12. Let $V \in \mathbb{R}^2$, T be a reflection transformation over the Y axis, i.e. $T(x,y) = (-x,y)$ and S are subspaces of V, $S = \{(x,y) \in \mathbb{R}^2 \mid x,y \in \mathbb{R}, y = 0\}$. (i) Is S a subspace of V? (ii) Is S an invariant subspace of V. (iii) If $S_1 = \{(x,y) \in \mathbb{R}^2 \mid x,y \in \mathbb{R}, x = 1, y = 1\}$, is S_1 an invariant subspace of V?.

Solution 2.11. (i) Yes, it is. (a) $x = 0 \Rightarrow (0,0) \in S$, (b) Let $\forall x_1, x_2 \in \mathbb{R}$, $(x_1, 0) + (x_2, 0) = (x_1 + x_2, 0) \in S$ and (c) $\alpha \in \mathbb{R}$, $(\alpha x, 0) \in S$. So S is a subspace of V (Def. 5). (ii) Yes, it is. $\forall x \in \mathbb{R}$, $(x, 0) \in S$, $T(x, 0) = (-x, 0) \in S$ then $T(x, y) \subseteq S$. (iii) No, it is not. $T(x_1, x_1) = (-x_1, x_1) \notin S$. So S_1 is not an invariant subspace of V.

Example 2.13. Let $V \in \mathbb{R}^3$, $T(x,y,z)$ be a rotation transformation over the Z axis

$$T(x,y,z) = \begin{pmatrix} \cos\theta & -\sin\theta & 0 \\ \sin\theta & \cos\theta & 0 \\ 0 & 0 & 1 \end{pmatrix} \begin{pmatrix} x \\ y \\ z \end{pmatrix}_{\theta=90°} = \begin{pmatrix} 0 & -1 & 0 \\ 1 & 0 & 0 \\ 0 & 0 & 1 \end{pmatrix} \begin{pmatrix} x \\ y \\ z \end{pmatrix} = \begin{pmatrix} -y \\ x \\ z \end{pmatrix}$$

and S a subspace of V, $S = \{(x,y,z) \in \mathbb{R}^3 \mid x,y,z \in \mathbb{R}, z = 0\}$. (i) Is S a subspace of V? (ii) Is S an invariant subspace of V?

Solution 2.12. (i) Yes, it is. (a) $x = 0, y = 0, z = 0 \Rightarrow (0,0,0) \in S$, (b) Let $\forall x_1, y_1, x_2, y_2 \in \mathbb{R}$, $(x_1, y_1, 0) + (x_2, y_2, 0) = (-y1 - y_2, x_1 + x_2, 0) \in S$, and (c) $\alpha \in \mathbb{R}$, $\alpha(x, y, 0) \in S$. So S is a subspace of V. (Def. 5). (ii) Yes, it is. $\forall x, y, z \in \mathbb{R}$, $(x, y, 0) \in S$, $T(x, y, 0) = (-y, x, 0) \in S$ then $T(x, y, z) \subseteq S$.

Note 2.5. "invariant" means $\forall s \in T(s)$, $T(s) \subseteq S$, **it does not mean** $T(s) = S$.

CHAPTER 3

Nonlinear Transformations

Carlos Polanco[*]

Faculty of Sciences, Universidad Nacional Autónoma de México, México

Abstract: This chapter introduces the definition of nonlinear transformations, their basic properties, and components. It explains their main applications through different examples and gives a first classification, their restrictions, and extensions. It includes special cases showing that it is possible to transform nonlinear equations into linear equations, using nonlinear transformations and it also describes the Least Squares method.

Keywords: Double homothetic factor, Least squares method, Linear transformation, Linearization, Logarithmic model, Nonlinear transformation, Reciprocal model.

3.1. NONLINEAR TRANSFORMATIONS

Definition 3.1. Suppose A and B are linear spaces over a field F. A transformation $T: A \to B$ is said to be **nonlinear** if $\exists\ x, y \in A$ or $\exists\ \alpha \in F$ [Carrell, 2005] Ch. 6, where any of these conditions (3.1, 3.2) are met:

$$T(x + y) \neq T(x) + T(y) \tag{3.1}$$

$$\alpha T(x) \neq T(\alpha x) \tag{3.2}$$

3.1.1. Examples of Nonlinear Transformations

$\alpha \neq 0$, *i.e.*, if all the plane moves to a fixed direction α, it will result in a nonlinear transformation. Thus, it will be easy to proof that a translation is not a linear function.

Example 3.1. Be a translation transformation $T_1: \mathbb{R}^2 \to \mathbb{R}^2$; $(x, y) \to (x + 1, y + 0)$. Is T_1 nonlinear? (Example 1.19).

---

[*]**Corresponding author Carlos Polanco:** Faculty of Sciences, Universidad Nacional Autónoma de México, México City, México; Tel: +01 55 5622 4858; Fax: +01 5556 4859; E-mail: polanco@unam.mx

Solution 3.1. Let $T_1(0,0) = (0 + 1,0) = (1,0) \neq (0,0)$. T_1 is nonlinear (Note 2.1).

Example 3.2. Be the translation transformation $T_1 : \mathbb{R}^2 \to \mathbb{R}^2$; $(x,y) \to (x + 1, y + 0)$, and $T_2 : \mathbb{R}^2 \to \mathbb{R}^2$; $(x,y) \to (2x, 2y)$ the "double" homothety factor. (i) Is $T_1 \circ T_2$ nonlinear? (ii) Is $T_2 \circ T_1$ nonlinear?

Solution 3.2. (i) Let $T_1 \circ T_2(x,y) = T_1(2x, 2y) = (2x + 1, 2y)$. Clearly $T_1 \circ T_2(0,0) = (0 + 1,0) = (1,0) \neq (0,0)$. $T_1 \circ T_2$ is nonlinear. (ii) Let $T_2 \circ T_1(x,y) = T_2(x + 1, y) = (2(x + 1), 2y) = (2x + 2, 2y)$. Then $T_2 \circ T_1(0,0) = (2,0) \neq (0,0)$. $T_1 \circ T_2$ is nonlinear.

Example 3.3. Be transformation $T_1 : [\mathbb{R}^+ \times \mathbb{R}^+] \to \mathbb{R}^2$; $(x,y) \to (\sqrt{x}, \sqrt{y})$ and $T_2 : [\mathbb{R}^+ \times \mathbb{R}^+] \to \mathbb{R}^2$; $(x,y) \to (x^2, y^2)$. (i) Is $T_1 \circ T_2$ nonlinear? (ii) Is $T_2 \circ T_1$ nonlinear? (iii) Is T_1 linear? (iv)What are your conclusions?

Solution 3.3. (i) Clearly $T_1 \circ T_2(x,y) = (x,y)$. So this composition is linear. (ii) Let $T_2 \circ T_1(x,y) = (x,y)$. So T is also linear. (iii) $\alpha T_1(x,y) = (\alpha\sqrt{x}, \alpha\sqrt{y}) \neq T_1(\alpha x, \alpha y) = (\sqrt{\alpha x}, \sqrt{\alpha y})$, $T_1(x_1 + x_2, y_1 + y_2) = (\sqrt{x_1 + x_2}, \sqrt{y_1 + y_2}) \neq T_1(x_1, y_1) + T_1(x_2, y_2)$. Therefore, T_1 is nonlinear.

3.2. LINEARIZATION

3.2.1. Logarithmic Model

A family of **nonlinear equations** can be transformed into **linear equations** by using the auxiliary nonlinear transformation $T(x) = \log x$ [Kardi, 2016].

3.2.1.1. Case: $T_1(x) = \alpha\beta^x$

Be $T_1 : \mathbb{R} \to \mathbb{R}$: $x \to \alpha\beta^x$; $\alpha, \beta \in \mathbb{R}^+$, and $T : \mathbb{R} \to \mathbb{R}$: $x \to \log x \Rightarrow T \circ T_1(x) = \log(\alpha\beta^x) = \log\alpha + \log(\beta^x) = \log\alpha + x\log\beta$. (i) Is T_1 a nonlinear transformation? (ii) Is $T \circ T_1(x)$ a **nonlinear transformation**? (iii) Is T a nonlinear transformation? (iv)What is the inverse transformation? (v) Is the inverse transformation a nonlinear transformation? (vi) Is $T \circ T_1(x)$ transformation representing a **linear equation**?

Proof. (i) $T_1(x_1 + x_2) = \alpha\beta^{x_1 + x_2} = \alpha\beta^{x_1}\beta^{x_2} \neq \alpha\beta^{x_1} + \alpha\beta^{x_2} = T_1(x_1) + T_1(x_2)$. So T_1 is a nonlinear transformation. (ii) Be $T \circ T_1(x) = \log\alpha + x\log\beta \Rightarrow T \circ T_1(x_1 + x_2) = \log\alpha + (x_1 + x_2)\log\beta = \log\alpha + x_1\log\beta + x_2\log\beta \neq \log\alpha + x_1\log\beta + \log\alpha + x_2\log\beta = T \circ T_1(x_1) + T \circ T_1(x_2)$. So $T \circ T_1(x)$ is a nonlinear transformation. (iii) $T(x_1 + x_2) = \log(x_1 + x_2) \neq \log x_1 + \log x_2 = \log(x_1 x_2)$. So

T is a nonlinear transformation. (iv) In general $T(x) = \log_b(x) \Rightarrow T^{-1}(x) = b^x$, because $T \circ T^{-1}(x) = \log_b(b^x) = id(x) = x$, $\log_a a = 1$.   (v)   Yes,   it   is. Particularly if $T^{-1}(x) = e^x \Rightarrow T^{-1}(x_1 + x_2) = e^{x_1+x_2} = e^{x_1}e^{x_2} \neq e^{x_1} + e^{x_2} = T^{-1}(x_1) + T^{-1}(x_2)$. So T^{-1} is a nonlinear transformation. (vi) Yes, the equation $T \circ T_1(x) = \log\alpha + x\log\beta$ is representing a **linear equation**.

Note 3.1. In general, an f function where $f(x) = mx + b$, $m, b \in \mathbb{R}$ is not necessarily a **linear transformation**. Instead, it is a **scalar-valued function** *i.e.*, f is a **linear equation**.

Note 3.2. It is advisable to use logarithmic graph paper to graphically represent the $\log T_2(x) = \log\alpha + \beta\log x$.

3.2.1.2. Case: $T_2(x) = \alpha x^\beta$

Be $T_2: \mathbb{R} \to \mathbb{R}: x \to \alpha x^\beta$; $\alpha, \beta \in \mathbb{R}^+$, and $T: \mathbb{R} \to \mathbb{R}: x \to \log x \Rightarrow T \circ T_2(x) = \log(\alpha x^\beta) = \log\alpha + \log(x^\beta) = \log\alpha + \beta\log x$. (i) Is T_2 a nonlinear transformation? (ii) Is $T \circ T_2(x)$ a **nonlinear transformation**? (iii) Is T a nonlinear transformation? (iv) What is the inverse transformation? (v) Is the inverse transformation a nonlinear transformation? (vi) Is $T \circ T_2(x)$ transformation representing a **linear equation**?

Proof. (i) $T_2(x_1 + x_2) = \alpha(x_1 + x_2)^\beta \neq T_2(x_1) + T_2(x_2) = \alpha x_1^\beta + \alpha x_2^\beta$. If $\rho \in \mathbb{R}$ $\Rightarrow \rho T_2(x) = \rho\alpha x^\beta \neq T_2(\rho x) = \alpha(\rho x)^\beta$. So T_2 is a nonlinear transformation. (ii) Be $T \circ T_2(x) = \log\alpha + \beta\log x \Rightarrow T \circ T_2(x_1 + x_2) = \log\alpha + \beta\log(x_1 + x_2) \neq \log\alpha + \beta\log x_1 + \log\alpha + \beta\log x_2 = T \circ T_2(x_1) + T \circ T_2(x_2)$. So $T \circ T_2(x)$ is a nonlinear transformation. (iii) $T(x_1 + x_2) = \log(x_1 + x_2) \neq \log x_1 + \log x_2 = \log(x_1 x_2)$. So T is a nonlinear transformation. (iv) In general $T(x) = \log_b(x) \Rightarrow T^{-1}(x) = b^x$, because $T \circ T^{-1}(x) = \log_b(b^x) = id(x) = x$, $\log_a a = 1$. (v) Yes it is. Particularly if $T^{-1}(x) = e^x \Rightarrow T^{-1}(x_1 + x_2) = e^{x_1+x_2} = e^{x_1}e^{x_2} \neq e^{x_1} + e^{x_2} = T^{-1}(x_1) + T^{-1}(x_2)$. So T^{-1} is a nonlinear transformation. (vi) Yes, the equation $T \circ T_2(x) = \log\alpha + \beta\log x$ is representing a **linear equation**.

Note 3.3. In general, an f function where $f(x) = mx + b$, $m, b \in \mathbb{R}$ is not necessarily a **linear transformation**. Instead, it is a **scalar-valued function** *i.e.*, f is a **linear equation**.

Note 3.4. It is advisable to use logarithmic graph paper to graphically represent the $\log T_2(x) = \log\alpha + \beta\log x$.

3.2.1.3. Case: $T_3(x) = \alpha e^{\beta x}$

Be $T_3 \colon \mathbb{R} \to \mathbb{R} \colon x \to \alpha e^{\beta x}$; $\alpha, \beta \in \mathbb{R}^+$, $e = $ constant, and $T \colon \mathbb{R} \to \mathbb{R} \colon x \to$ $\ln x \Rightarrow T \circ T_3(x) = \ln(\alpha e^{\beta x}) = \ln\alpha + \ln(e^x \beta x) = \ln\alpha + \beta x \ln e = \ln\alpha + \beta x$. (i) Is T_3 a nonlinear transformation? (ii) Is $T \circ T_3(x)$ a **nonlinear transformation**? (iii) Is T a nonlinear transformation? (iv)What is the inverse transformation? (v) Is the inverse transformation a nonlinear transformation? (vi) Is $T \circ T_3(x)$ transformation representing a **linear equation**?

Proof. (i) $T_3(x_1 + x_2) = \alpha e^{\beta(x_1+x_2)} = \alpha e^{\beta x_1} e^{\beta x_2} \neq \alpha e^{\beta x_1} + \alpha e^{\beta x_2} = T_3(x_1) + T_3(x_2)$. So T_3 is a nonlinear transformation. (ii) Be $T \circ T_3(x) = \ln\alpha + \beta x \Rightarrow T \circ T_3(x_1 + x_2) = \ln\alpha + \beta(x_1 + x_2) = \ln\alpha + \beta x_1 + x_2 \beta x_2 \neq \ln\alpha + \beta x_1 + \ln\alpha + \beta x_2 = T \circ T_3(x_1) + T \circ T_3(x_2)$. So $T \circ T_3(x)$ is a nonlinear transformation. (iii) $T(x_1 + x_2) = \ln(x_1 + x_2) \neq \ln x_1 + \ln x_2 = \ln(x_1 x_2)$. So T is a nonlinear transformation. (iv) In general, $T(x) = \ln_b(x) \Rightarrow T^{-1}(x) = b^x$, because $T \circ T^{-1}(x) = \ln_b(b^x) = id(x) = x$, $\ln e = 1$. (v) Yes, it is. Particularly if $T^{-1}(x) = e^x \Rightarrow T^{-1}(x_1 + x_2) = e^{x_1+x_2} = e^{x_1} e^{x_2} \neq e^{x_1} + e^{x_2} = T^{-1}(x_1) + T^{-1}(x_2)$. So T^{-1} is a nonlinear transformation. (vi) Yes, the equation $T \circ T_3(x) = \ln\alpha + \beta x$ is representing a **linear equation**.

Note 3.5. In general, an f function where $f(x) = mx + b$, $m, b \in \mathbb{R}$ is not necessarily a **linear transformation**. Instead, it is a **scalar-valued function** *i.e.*, f is a **linear equation**.

Note 3.6. It is advisable to use logarithmic graph paper to graphically represent $\ln T_3(x) = \ln\alpha + \beta x$.

3.2.1.4. Case: $T_4(x) = \ln\alpha x^{\beta}$

Be $T_4 \colon \mathbb{R} \to \mathbb{R} \colon x \to \ln\alpha x^{\beta}$; $\alpha, \beta \in \mathbb{R}^+$, and $T \colon \mathbb{R} \to \mathbb{R} \colon x \to \ln x \Rightarrow T \circ T_4(x) = \ln\alpha x^{\beta} = \ln\alpha + \ln(x^{\beta}) = \ln\alpha + \beta \ln x$, or $T_4(x) = \ln\alpha x^{\beta} \Leftrightarrow e^{T_4(x)} = e^{\ln\alpha x^{\beta}} \Leftrightarrow \ln e^{T_4(x)} = T_4(x) = \ln\alpha + \beta \ln x$. (i) Is T_4 a nonlinear transformation? (ii) Is $T \circ T_4(x)$ a **nonlinear transformation**? (iii) Is T a nonlinear transformation? (iv) What is the inverse transformation? (v) Is the inverse transformation a nonlinear transformation? (vi) Is $T \circ T_4(x)$ transformation representing a **linear equation**?

Proof. (i) $T_4(x_1 + x_2) = \ln\alpha + \beta \ln x_1 + x_2 = \alpha + \beta(\ln x_1 \ln x_2) \neq \alpha + \beta \ln x_1 + \ln\alpha + \beta \ln x_2 = T_4(x_1) + T_4(x_2)$. So T_4 is a nonlinear transformation. (ii) Be $T \circ T_4(x) = \ln\alpha + \beta \ln x \Rightarrow T \circ T_4(x_1 + x_2) = \ln\alpha + \beta \ln(x_1 + x_2) = \ln\alpha + \beta \ln x_1 +$

$\beta \ln x_2 \neq \ln \alpha + \beta \ln x_1 + \ln \alpha + \beta \ln x_2 = T \circ T_4(x_1) + T \circ T_4(x_2)$. So $T \circ T_4(x)$ is a nonlinear transformation. (iii) $T(x_1 + x_2) = \ln(x_1 + x_2) \neq \ln x_1 + \ln x_2 = \ln(x_1 x_2)$. So T is a nonlinear transformation. (iv) In general, $T(x) = \ln_b(x) \Rightarrow T^{-1}(x) = b^x$, because $T \circ T^{-1}(x) = \ln_b(b^x) = id(x) = x$, $\ln_a a = 1$. (v) Yes, it is. Particularly if $T^{-1}(x) = e^x \Rightarrow T^{-1}(x_1 + x_2) = e^{x_1 + x_2} = e^{x_1} e^{x_2} \neq e^{x_1} + e^{x_2} = T^{-1}(x_1) + T^{-1}(x_2)$. So T^{-1} is a nonlinear transformation. (vi) Yes, the equation $T \circ T_4(x) = \ln \alpha + \beta \ln x$ is representing a **linear equation**.

Note 3.7. In general, an f function where $f(x) = mx + b$, $m, b \in \mathbb{R}$ is not necessarily a **linear transformation**. Instead, it is a **scalar-valued function** *i.e.*, f is a **linear equation**.

Note 3.8. It is advisable to use logarithmic graph paper to graphically represent $\ln T_3(x) = \ln \alpha + \beta x$.

3.2.2. Reciprocal Model

A family of **nonlinear equations** can be transformed into **linear equations** by using the auxiliary nonlinear transformation $T(x) = \frac{1}{x}$ [Kardi, 2016].

3.2.2.1. Case: $T_5(x) = \frac{1}{\alpha + \beta x}$

Be $T_5 \colon \mathbb{R} \to \mathbb{R} \colon x \to \frac{1}{\alpha + \beta x}$; $\alpha, \beta \in \mathbb{R}$, and $T \colon \mathbb{R} \to \mathbb{R} \colon x \to \frac{1}{x} \Rightarrow T \circ T_5(x) = \frac{1}{T_5(x)} = \alpha + \beta x$. (i) Is T_5 a nonlinear transformation? (ii) Is $T \circ T_5(x)$ a **nonlinear transformation**? (iii) Is T a nonlinear transformation? (iv) What is the inverse transformation? (v) Is the inverse transformation a nonlinear transformation? (vi) Is $T \circ T_5(x)$ transformation representing a **linear equation**?

Proof. (i) $T_5(x_1 + x_2) = \frac{1}{\alpha + \beta(x_1 + x_2)} \neq \frac{1}{\alpha + \beta x_1} + \frac{1}{\alpha + \beta x_2} = T_5(x_1) + T_5(x_2)$. So T_5 is a nonlinear transformation. (ii) Be $T \circ T_5(x) = \frac{1}{\alpha + \beta x} \Rightarrow T \circ T_5(x_1 + x_2) = \alpha + \beta(x_1 + x_2) \neq \alpha + \beta x_1 + \alpha + \beta(x_2 = T \circ T_5(x_1) + T \circ T_5(x_2)$. So $T \circ T_5(x)$ is a nonlinear transformation. (iii) $T(x_1 + x_2) = \frac{1}{x_1 + x_2} \neq T(x_1) + T(x_2)$. So T is a nonlinear transformation. (iv) $T^{-1}(x) = x$, because $T \circ T^{-1}(x) = \frac{1}{\frac{1}{x}} = id(x) = x$. (v) Yes, it is. If $T^{-1}(x) = x \Rightarrow T^{-1}(x_1 + x_2) = x_1 + x_2 == x_1 + x_2 = T^{-1}(x_1) + T^{-1}(x_2)$ and $T^{-1}(\alpha x) = \alpha T^{-1}(x)$. So T^{-1} is a **linear transformation**.

(vi) Yes, the equation $T \circ T_5(x) = \frac{1}{T_5(x)} = \alpha + \beta x$ is representing a **linear equation**.

Note 3.9. Note that in the graph, the Y-axis represents $\frac{1}{T_5(x)}$.

3.2.3. Square Root Model

A family of **nonlinear equations** can be transformed into **linear equations** by using the auxiliary nonlinear transformation $T(x) = \sqrt{x}$ [Kardi, 2016].

3.2.3.1. Case: $T_6(x) = \frac{1}{(\alpha+\beta x)^2}$

Be $\quad T_6 : \mathbb{R} \to \mathbb{R} : x \to \frac{1}{(\alpha+\beta x)^2}$; $\alpha, \beta \in \mathbb{R}^+$, and $T : \mathbb{R} \to \mathbb{R} : x \to \sqrt{x} \Rightarrow T \circ T_6(x) = \sqrt{T_6(x)} = \frac{1}{\sqrt{(\alpha+\beta x)^2}}$. (i) Is T_6 a nonlinear transformation? (ii) Is $T \circ T_6(x)$ a **nonlinear transformation**? (iii) Is T a nonlinear transformation? (iv) What is the inverse transformation? (v) Is the inverse transformation a nonlinear transformation? (vi) Is $T \circ T_6(x)$ transformation representing a **linear equation**?

Proof. (i) $T_6(x_1 + x_2) = \frac{1}{(\alpha+\beta(x_1+x_2))^2} \neq \frac{1}{(\alpha+\beta x_1)^2} + \frac{1}{(\alpha+\beta x_2)^2} = T_6(x_1) + T_6(x_2)$. So T_6 is a nonlinear transformation. (ii) Be $T \circ T_6(x) = \sqrt{T_6(x)} = \frac{1}{\sqrt{(\alpha+\beta x)^2}} = \frac{1}{\alpha+\beta x} \Rightarrow T \circ T_6(x_1 + x_2) = \alpha + \beta(x_1 + x_2) \neq \alpha + \beta x_1 + \alpha + \beta(x_2 = T \circ T_6(x_1) + T \circ T_6(x_2)$. So $T \circ T_6(x)$ is a nonlinear transformation. (iii) $T(x_1 + x_2) = \sqrt{x_1 + x_2} \neq T(x_1) + T(x_2)$. So T is a nonlinear transformation. (iv) $T^{-1}(x) = x$; $x \in \mathbb{R}^x$, because $T \circ T^{-1}(x) = \sqrt{x^2} = id(x) = x$. (v) Yes, it is. If $T^{-1}(x) = x^2 \Rightarrow T^{-1}(x_1 + x_2) = (x_1 + x_2)^2 \neq x_1^2 + x_2^2 = T^{-1}(x_1) + T^{-1}(x_2)$. So T^{-1} is a nonlinear transformation. (vi) Yes, the equation $T \circ T_6(x) = \sqrt{T_6(x)} = \frac{1}{\sqrt{(\alpha+\beta x)^2}}$ is representing a **linear equation**.

Note 3.10. Note that in the graph the Y axis represents $\sqrt{T_6(x)}$.

3.2.4. Least Squares Method

This method builds a linear equation from a succession of n data points *i.e.*, $(x_1, y_1), (x_2, y_2), \cdots, x_n, y_n) \in \mathbb{R}^2$, minimizing the distance from these data points to the proposed linear equation. This method [Spiegel, 2001] Ch. 13 provides an assessment of the quality of this approach, through the **Pearson Correlation Coefficient** [Spiegel, 2001] Ch. 14. If r tends to -1 or 1 the data points lie exactly on a line. Once verified r, the linear equation is calculated using the **Least Squares method**.

$$r = \frac{n\sum_{i=1}^n x_i y_i - \sum_{i=1}^n x_i \sum_{i=1}^n y_i}{\sqrt{[n\sum_{i=1}^n x_i^2 - (\sum_{i=1}^n x_i)^2]\ [n\sum_{i=1}^n y_i^2 - (\sum_{i=1}^n y_i)^2]}} \in [-1,1]. \tag{3.3}$$

The linear equation $T(x) = a_0 + a_1 x$ where $a_0, a_1 \in \mathbb{R}$, is built from the solution of the system of equations.

$$\sum_{i=1}^n y_i = a_0 n \quad + a_1 \sum_{i=1}^n x_i$$

$$\sum_{i=1}^n x_i y_i = a_0 \sum_{i=1}^n x_i + a_1 \sum_{i=1}^n x_i^2$$

$$a_0 = \frac{\sum_{i=1}^n y_i \sum_{i=1}^n x_i^2 - \sum_{i=1}^n x_i \sum_{i=1}^n x_i y_i}{n\sum_{i=1}^n x_i^2 - (\sum_{i=1}^n x_i)^2}, a_1 = \frac{n\sum_{i=1}^n x_i y_i - \sum_{i=1}^n x_i \sum_{i=1}^n y_i}{n\sum_{i=1}^n x_i^2 - (\sum_{i=1}^n x_i)^2}$$

3.2.5. Example of Nonlinear Equation

Example 3.4. Table **3.1** shows the experimental values between pressure P an volume V. (i) Verify the relation $PV^\rho = C$, where ρ and C are real constants. (ii) Calculate the Pearson Correlation Coefficient. (iii) Estimate the values of ρ and C. (iv) Obtain the equation that relates P and V. (v) Estimate P if $V = 93.1\ pulg^3$.

Table 3.1. Volume *versus* temperature.

Volume V ($pulg^3$)	54.6	62.1	72.7	89.0	118.9	194.3
Pressure P ($lb/pulg^2$)	61.5	49.5	37.9	28.7	19.5	10.4

Experimental values of volume V and temperature T. Source: [Spiegel, 2001] Ch. 6, Ex. 13.21.

Solution 3.4. (i) If $PV^\rho = C$ is true, then $\log_{10} P \rho \log_{10} V = \log_{10} C \Leftrightarrow \log_{10} P = \log_{10} C - \rho \log_{10} V$. We substitute $X = \log_{10} V$ and $Y = \log_{10} P \Rightarrow Y = a_0 + a_1 X$, where $a_0 = \log_{10} C$, and $a_1 = -\rho$.

Table 3.2. Pearson correlation coefficient.

$X = \log_{10}V$	$Y = \log_{10}P$	X^2	Y^2	XY
1.737	1.789	3.017	3.201	3.107
1.793	1.695	3.215	2.873	3.039
1.862	1.579	3.467	2.493	2.940
1.949	1.458	3.799	2.126	2.842
2.075	1.290	4.306	1.664	2.677
2.288	1.017	5.235	1.034	2.327
$\sum_{i=1}^{6} = 11.704$	$\sum_{i=1}^{6} = 8.828$	$\sum_{i=1}^{6} = 23.039$	$\sum_{i=1}^{6} = 13.391$	$\sum_{i=1}^{6} = 16.932$

Variables to calculate the Pearson Correlation Coefficient.

$$r = \frac{n\sum_{i=1}^{n} x_i y_i - \sum_{i=1}^{n} x_i \sum_{i=1}^{n} y_i}{\sqrt{[n\sum_{i=1}^{n} x_i^2 - (\sum_{i=1}^{n} x_i)^2]\ [n\sum_{i=1}^{n} y_i^2 - (\sum_{i=1}^{n} y_i)^2]}}$$

$$= \frac{6(16.932) - (11.704)(8.828)}{\sqrt{[6(23.039) - (11.704)^2][6(13.391) - (8.828)^2]}}$$

$$= -0.99637$$

(ii) So $\log_{10}P = \log_{10}C - \rho\log_{10}V$ is a linear equation. We assign $X = \log_{10}V$ and $Y = \log_{10}P \Rightarrow Y = a_0 + a_1 X$. According to

$$a_0 = \frac{\sum_{i=1}^{n} y_i \sum_{i=1}^{n} x_i^2 - \sum_{i=1}^{n} x_i \sum_{i=1}^{n} x_i y_i}{n\sum_{i=1}^{n} x_i^2 - (\sum_{i=1}^{n} x_i)^2} = \frac{8.828(23.039) - (11.704)(16.932)}{6(23.039) - (11.704)^2}$$

$$= 4.1716$$

$$a_1 = \frac{n\sum_{i=1}^{n} x_i y_i - \sum_{i=1}^{n} x_i \sum_{i=1}^{n} y_i}{n\sum_{i=1}^{n} x_i^2 - (\sum_{i=1}^{n} x_i)^2} = \frac{6(16.932) - (11.704)(8.828)}{6(23.039) - (11.704)^2}$$

$$= -1.3843$$

So $Y = 4.1716 - 1.3843X$. (iii) Since $a_0 = 4.1716 = \log_{10}C$ and $a_1 = -1.3843 = -\rho \Rightarrow \rho = 1.3843$ and $C = 1.4846 \times 10^4$. (iv) The equation will be

$PV^{1.3843} = 1.4846$. (v) If $V = 93.1 \ pulg^3 \rightarrow X = \log_{10}V = 1.969$ and $Y = \log_{10}P = 4.1716 - 1.3843(1.969) = 1.4459 \Rightarrow P = 10^{1.4459} = 27.919 lb/pulg^2$.

3.3. LOGARITHMIC FUNCTION

Definition 3.2. Let us take the exponential function $f(x) = b^x$, where $b \in \mathbb{R}$, $f^{-1}(x) = \log_b x, \Rightarrow f \circ f^{-1}(x) = \text{Id}(x) = x$ *i.e.*, $f^{-1} \circ f(x) = f^{-1}(\log_b x) = b^{\log_b x} = x$. The $f^{-1}(x)$ function is known as logarithmic function, *e.g.*, $f(x) = 3^x \Rightarrow f^{-1}(x) = \log_3 x \Rightarrow f^{-1} \circ f(x) = f^{-1}(\log_3 x) = 3^{\log_3 x} = x$ [Wikipedia, 2017d].

3.3.1. Properties

For any base b, $\log_b b = 1$ and $\log_b 1 = 0$, since $b^1 = b$, and $b^0 = 1$, respectively, there are five properties that have to be verified: (i) $\log_b xy = \log_b x + \log_b y$. (ii) $\log_b \left(\frac{x}{y}\right) = \log_b x - \log_b y$. (iii) $\log_b x^p = p\log_b x$. (iv) $\log_b \sqrt[p]{x} = \frac{\log_b x}{p}$. (v) $\log_a x = \frac{\log_b x}{\log_b a}$. "The identity $\log xy = \log x + \log d$ expresses a group isomorphism between positive reals under multiplication and reals under addition. Logarithmic functions are the only continuous isomorphisms between these groups" [Bourbaki, 1998; Wikipedia, 2017d] §$V.4.1$.

Example 3.5. Solve the following exercises (i) $\log_{10} 100 = y$. (ii) $\log_{10} 0.0001$. (iii) $\log_{10} \sqrt[4]{1000000}$. (iv) $\ln\sqrt{e}$. (v) $\log_6 4 + \log_6 9$.

Solution 3.5. (i) $10^{\log_{10} 100} = 10^y \Leftrightarrow 10^{\log_{10} 10^2} = 10^y \Leftrightarrow 10^2 = 10^y \Rightarrow y = 2$.
(ii) $\log_{10} 0.0001 = y \Leftrightarrow 10^{\log_{10} 0.00001} = 10^y \Leftrightarrow 10^{\log_{10} 10^{-4}} = 10^y \Leftrightarrow 10^{-4} = 10^y \Rightarrow y = -4$.

Note 3.11. $0.00001 = 10^{-2}10^{-2} \Rightarrow \log 10^{-4} = \log_{10} 10^{-2} + \log_{10} 10^{-2}$ then $y = \log_{10} 10^{-2} + \log_{10} 10^{-2} = -2 - 2 = -4$.

(iii)   If $y = \log_{10} \sqrt[4]{1000000} \Rightarrow 10^y = 10^{\log_{10} \sqrt[4]{1000000}} = 10^{\log_{10} \sqrt[4]{10^6}} = 10^{\log_{10} 10^{6/4}} = 10^{6/4} \Rightarrow y = 6/4 = 3/2$. (iv) $y = \ln\sqrt{e} \Leftrightarrow e^y = e^{\ln\sqrt{e}} \Leftrightarrow e^y = e^{1/2} \Rightarrow y = 1/2$. (v) $y = \log_6 4 + \log_6 9 \Leftrightarrow y = \log_6 36 = 2$.

Example 3.6. Be $y(x) = \log_b x$. Is y a linear transformation?

Solution 3.6. $y(x_1 + x_2) = \log_b x_1 + x_2, \neq \log_b x_1 + \log_b x_2 = \log_b x_1 x_2$, in fact, $y(\alpha x) = \log_b \alpha x = \log_b \alpha + \log_b x \neq \alpha \log_b x$. So y is a nonlinear transformation.

3.3.2. Millimeter Scale

When the data Table **3.3** correlates with an exponential function, *e.g.*, $f(x) = \log_{10} x$, its plotting on millimeter scale (both axis), does not allow the proper appreciation of the small or large values Fig. (**3.1**). However, if this data is represented on logarithmic scale, it is possible to see a linear relationship Fig. (**3.2**) that is suitable for subsequent analysis.

Table 3.3 Data.

X — axis	1	2	3	4	5	6	7	8	9	10	11	12	20	40	50	100
Y — axis	0.00	0.30	0.47	0.60	0.69	0.77	0.84	0.90	0.95	1.00	1.04	1.07	1.30	1.60	1.69	2.00

Data obtained from function $f(x) = \log_{10} x$.

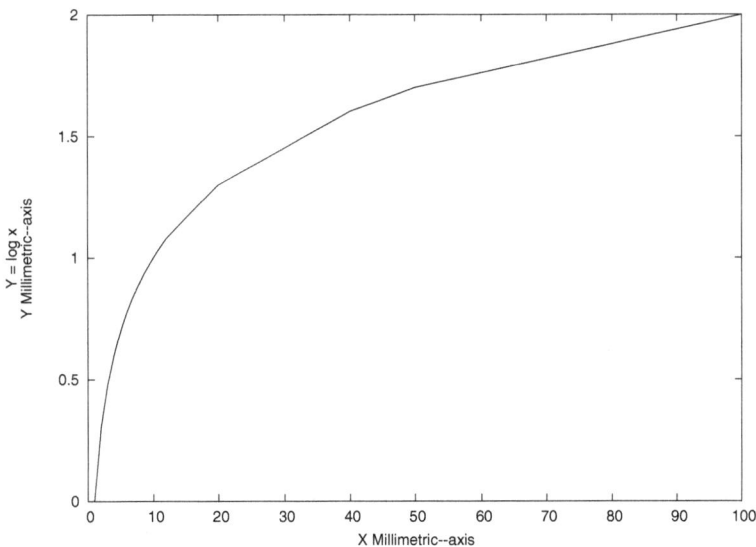

Fig. (3.1). Labeling a millimetric scale $f(x) = \log_{10} x$. Source Table **3.3**.

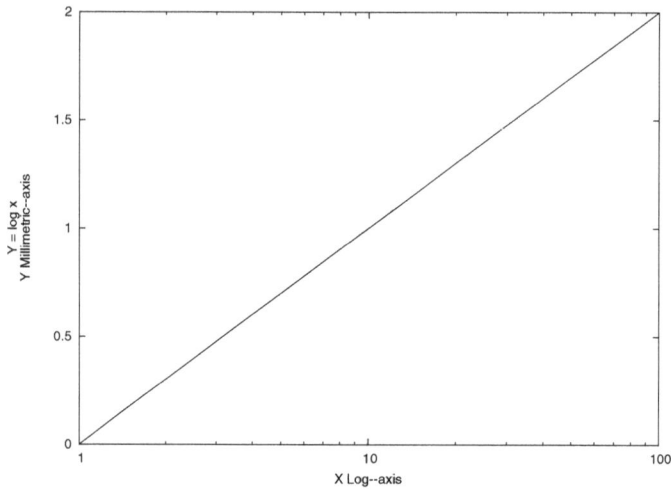

Fig. (3.2). Labeling a semi log scale $f(x) = \log_{10}x$. Source Table **3.3**.

3.3.3. Logarithmic Scale

Representing the data points on a logarithmic scale (Fig. **3.3**) is equivalent to represent the logarithms of these values on a millimeter scale. In these cases, it is very convenient to use logarithmic paper.

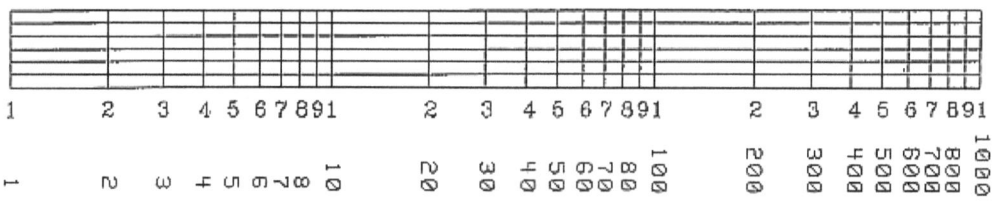

Fig. (3.3). Logarithmic scale. Courtesy of D. Simanek [Simanek, 2017].

CHAPTER 4

Acoustics

Carlos Polanco*

Faculty of Sciences, Universidad Nacional Autónoma de México, México

Abstract Acoustics is the branch of physics that studies all the physical phenomena associated with the generation, propagation, and detection of mechanical waves that are heard in a band of frequencies called sound waves.

Keywords: Fourier series.

4.1. BACKGROUND

Some real functions $f: \mathbb{R} \rightarrow \mathbb{R}$ can be represented through power series (Taylor series), which means they can be solved by polynomials. When periodicity is observed in the behavior of a function, the approximation with periodic functions becomes necessary *e.g.*, $\sin x$ or $\cos x$ [Abia, 2001]. This approximation is known as Fourier series.

4.2. FOURIER SERIES

This series (4.1) is attributed to the French mathematician Joseph Fourier (1768–1830) [Stewart, 1998], who was trying to solve a problem related to heat conduction. If $f(x)$ is a function whose integral is defined by $[-\frac{T}{2}, \frac{T}{2}]$, it is possible to obtain a Fourier series of f function in this interval. Outside this interval, the $f(x)$ function is periodic with period T. The approximation of $f(x)$ is

$$a_0 + \sum_{k=1}^{\infty} a_k \cos\left(\frac{2\pi k x}{T}\right) + b_k \sin\left(\frac{2\pi k x}{T}\right) \tag{4.1}$$

where

$$a_0 = \frac{1}{T} \int_{-\frac{T}{2}}^{\frac{T}{2}} f(x) \ dx,$$

---

*__Corresponding author Carlos Polanco:__ Faculty of Sciences, Universidad Nacional Autónoma de México, México City, México; Tel: +01 55 5622 4858; Fax: +01 5556 4859; E-mail: polanco@unam.mx

$$a_k = \frac{2}{T} \int_{-\frac{T}{2}}^{\frac{T}{2}} f(x) \cos(\frac{2\pi k x}{T}) \ dx,$$

and

$$b_k = \frac{2}{T} \int_{-\frac{T}{2}}^{\frac{T}{2}} f(x) \sin(\frac{2\pi k x}{T}) \ dx.$$

Example 4.1. Be the triangular wave function defined by [Stewart, 1998] $|x|, x \in [-\pi, \pi]$ and 0. (i) Calculate the Fourier series for $k = 1,2,3,4$, and 5. (ii) Is the Fourier series a linear transformation? (iii) Plot each approximation.

$$f(x) = \begin{cases} |x|: & x \in [-\pi, \pi] \\ 0 & : \ x \notin [-\pi, \pi] \end{cases}$$

Solution 4.1. (i) If $x \in [-\pi, \pi]$ and $k = 1 \Rightarrow T = 2\pi$. Then $a_0 = \frac{1}{2\pi} \int_{-\pi}^{\pi} |x| \ dx = \frac{\pi}{2}$; $a_1 = \frac{1}{\pi} \int_{-\pi}^{\pi} |x| \cos x \ dx = -\frac{4}{\pi}$; $b_1 = \frac{1}{\pi} \int_{-\pi}^{\pi} |x| \sin x \ dx = 0 \Rightarrow f_1(x) = a_0 + a_1 \cos x + b_1 \sin x = \frac{\pi}{2} - \frac{4}{\pi} \cos x$. If $k = 2 \Rightarrow a_2 = \frac{1}{\pi} \int_{-\pi}^{\pi} |x| \cos 2x \ dx = 0$; $b_2 = \frac{1}{\pi} \int_{-\pi}^{\pi} |x| \sin 2x \ dx = 0 \Rightarrow$ $f_2(x) = a_0 + \sum_{k=1}^{2} a_k \cos kx + b_k \sin kx = a_0 + a_1 \cos x + b_1 \sin x + a_2 \cos 2x + b_2 \sin 2x = \frac{\pi}{2} - \frac{4}{\pi} \cos x$. If $k = 3 \Rightarrow a_3 = \frac{1}{\pi} \int_{-\pi}^{\pi} |x| \cos 3x \ dx = -\frac{4}{9\pi}$; $b_3 = \frac{1}{\pi} \int_{-\pi}^{\pi} |x| \sin 3x \ dx = 0 \Rightarrow$ $f_3(x) = a_0 + \sum_{k=1}^{3} a_k \cos kx + b_k \sin kx = a_0 + a_1 \cos x + b_1 \sin x + a_2 \cos 2x + b_2 \sin 2x + a_3 \cos 3x + b_3 \sin 3x = \frac{\pi}{2} - \frac{4}{\pi} \cos x - \frac{4}{9\pi} \cos 3x$. If $k = 4 \Rightarrow a_4 = \frac{1}{\pi} \int_{-\pi}^{\pi} |x| \cos 4x \ dx = 0$; $b_4 = \frac{1}{\pi} \int_{-\pi}^{\pi} |x| \sin 4x \ dx = 0 \Rightarrow$ $f_4(x) = a_0 + \sum_{k=1}^{4} a_k \cos kx + b_k \sin kx = a_0 + a_1 \cos x + b_1 \sin x + a_2 \cos 2x + b_2 \sin 2x + a_3 \cos 3x + b_3 \sin 3x + a_4 \cos 4x + b_4 \sin 4x = \frac{\pi}{2} - \frac{4}{\pi} \cos x - \frac{4}{9\pi} \cos 3x$. If $k = 5 \Rightarrow a_5 = \frac{1}{\pi} \int_{-\pi}^{\pi} |x| \cos 5x \ dx = -\frac{4}{25\pi}$; $b_5 = \frac{1}{\pi} \int_{-\pi}^{\pi} |x| \sin 5x \ dx = 0 \Rightarrow f_5(x) = a_0 + \sum_{k=1}^{4} a_k \cos kx + b_k \sin kx = a_0 + a_1 \cos x + b_1 \sin x + a_2 \cos 2x + b_2 \sin 2x + a_3 \cos 3x + b_3 \sin 3x + a_4 \cos 4x + b_4 \sin 4x + a_5 \cos 5x + b_5 \sin 5x = \frac{\pi}{2} - \frac{4}{\pi} \cos x - \frac{4}{9\pi} \cos 3x - \frac{4}{25\pi} \cos 5x$. (ii) $f(\alpha x) = a_0 + \sum_{k=1}^{\infty} a_k \cos(\frac{2\pi k \alpha x}{T}) + b_k \sin(\frac{2\pi k \alpha x}{T}) \neq \alpha f(x)$. (ii) F(x) is a nonlinear transformation.

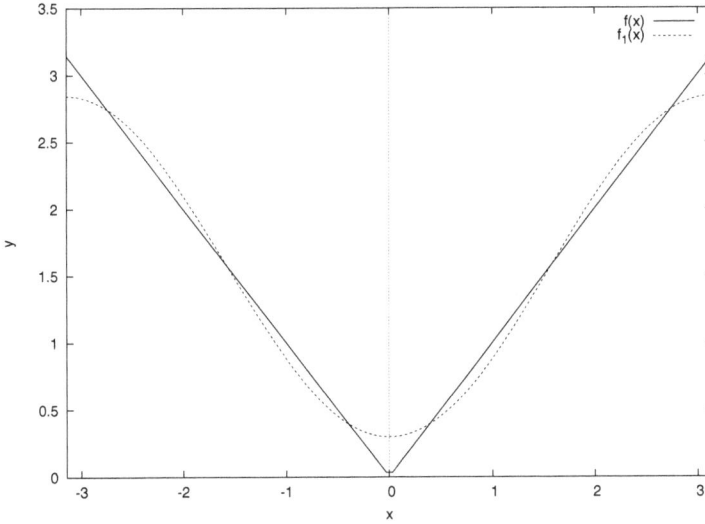

Fig. (4.1). Plot of functions $f(x) = |x|: x \in [-\pi, \pi], 0: x \notin [-\pi, \pi]$, and $f_1(x) = \frac{\pi}{2} - \frac{4}{\pi}\cos x$.

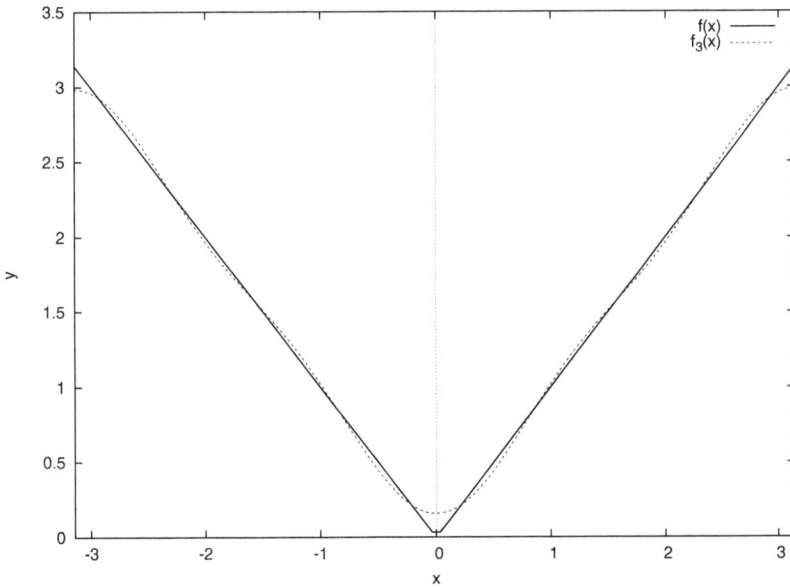

Fig. (4.2). Plot of functions $f(x) = |x|: x \in [-\pi, \pi], 0: x \notin [-\pi, \pi]$, and $f_3(x) = \frac{\pi}{2} - \frac{4}{\pi}\cos x - \frac{4}{9\pi}\cos 3x$.

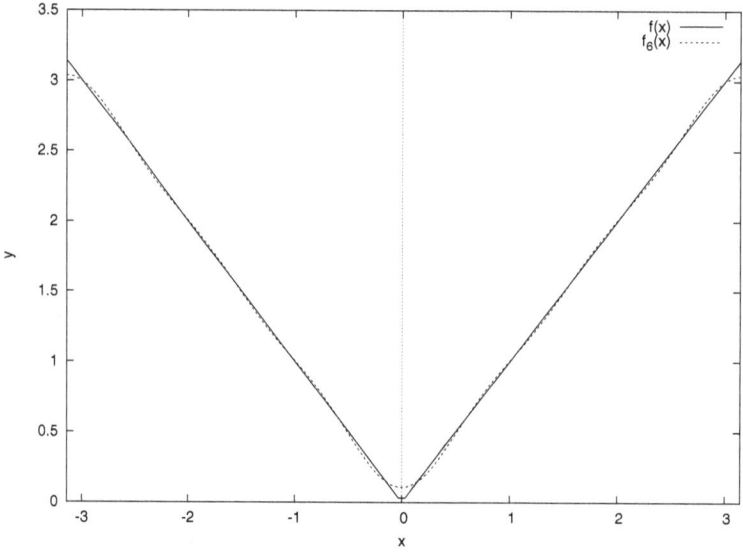

Fig. (4.3). Plot of functions $f(x) = |x|: x \in [-\pi, \pi], 0: x \notin [-\pi, \pi]$, and $f_5(x) = \frac{\pi}{2} - \frac{4}{\pi} \cos x - \frac{4}{9\pi} \cos 3x - \frac{4}{25\pi} \cos 5x$.

CHAPTER 5

Actuary

Carlos Polanco* and Dánae Itzel Álvarez

Faculty of Sciences, Universidad Nacional Autónoma de México, México

Abstract: Actuary is a discipline oriented to financial and commercial matters, where a mixture of algorithms of deterministic and stochastic nature comes together not only to predict the future value of specific parameters but also to estimate their past values. In this chapter, we review three issues: annuity, mortality rate, and linear programming.

Keywords: Annuity, Linear programming, Mortality rate.

5.1. BACKGROUND

5.2. ANNUITY

An annuity [Bowers *et al.*, 1997] is a series of equal payments made at equal intervals for a period of time. For instance, monthly rent payments, quarterly dividends on shares, monthly home mortgage payments, annual premium payments of a life insurance contract. The elements of an annuity are: the rent *i.e.*, the value of each periodic payment (R); the period of the rent *i.e.*, the interval between payments; the term of the annuity, *i.e.*, how long payments will be made (n); and the interest rate applied to each rent (i). The present value (P) of the annuity is defined as the total value of the rent payments at the beginning of the term for a period of time before the first payment. The equation to determine the value of the rent of the annuity is $R = P\left(\frac{i}{1-(1+i)^{-n}}\right)$, where the factor that multiplies the value is called the Cost of capital.

Example 5.1. A car is acquired with a loan of 60.000.00 USD to be paid in equal monthly payments, the debt is for a period of 2 years, with a monthly interest rate of 3%. (i) Determine the payments. (ii) Is the cost of capital equation a linear transformation?

---

*Corresponding author Carlos Polanco: Faculty of Sciences, Universidad Nacional Autónoma de México, México City, México; Tel: +01 55 5622 4858; Fax: +01 5556 4859; E-mail: polanco@unam.mx

Solution 5.1. (i) $R = P\left(\frac{i}{1-(1+i)^{-n}}\right) = 60,000\left(\frac{0.03}{1-(1+0.03)^{-24}}\right) = 60,000(0.059047)$

$= 3,542.84$ (ii) Considering equal interest rate and period of time, we have:

$T(x+y) = (x+y)\left(\frac{i}{1-(1+i)^{-n}}\right) = x\left(\frac{i}{1-(1+i)^{-n}}\right) + y\left(\frac{i}{1-(1+i)^{-n}}\right) = T(x) + T(y)$

and $T(\alpha x) = (\alpha x)\left(\frac{i}{1-(1+i)^{-n}}\right) = \alpha x\left(\frac{i}{1-(1+i)^{-n}}\right) = \alpha T(x)$. So the cost of capital

equation is a linear transformation.

5.3. MORTALITY RATE

Human biometrics is an actuarial statistics field that studies human survival and the concepts related to it, grouped in mortality tables. These mortality tables are very useful in areas such as demographics, insurance, actuarial science, and biostatistics, among others. With them, it is possible to analyze the behavior of the population to phenomena such as mortality, longevity, and fertility, to name a few.

The biometrics model is a stochastic model, its design is based on a random variable X, which is the age of death and represents the time gap between the birth of an individual and his death. One of the main functions of the biometrics model is the temporal probability of survival, which is denoted as $_tP_x$, and represents the probability an individual age x lives to age $x + t$.

It is possible to define $_tP_x$ (5.1) according to the mortality rate noted as $\mu(x)$, representing the mortality rate at age x for individuals that already reached that age. This is defined as

$$_tP_x = e^{-\int_x^{x+t}\mu(y)dy} \tag{5.1}$$

And, as it is illustrated below, this definition of the temporal probability of survival is a clear example of a non-linear transformation $_tP_x = e^{-\int_x^{x+t}\mu(y)dy}$ and $_tP_z = e^{-\int_z^{z+t}\mu(y)dy}$ So $_tP_x + _tP_z = e^{-\int_x^{x+t}\mu(y)dy} + e^{-\int_z^{z+t}\mu(y)dy}$. Thus $_tP_{x+z} = e^{-\int_{x+z}^{x+z+t}\mu(y)dy}$.

Example 5.2. Calculate the probability a person age 20 lives to age 25 given $\mu(x) = \frac{1}{100-x}$ con $0 \le x \le 100$.

Solution 5.2. $_5P_{20} = e^{-\int_{20}^{25}\mu(y)dy} = e^{-\int_{20}^{25}\frac{1}{100-x}dx} = e^{\ln(100-25)-\ln(100-20)} = e^{\ln(75)\ -\ln(80)} = 0.9375$

5.4. LINEAR PROGRAMMING

Linear programming [Hardy and Waters, 2009] is a mathematical modelling technique designed to optimize the use of limited resources. It is a useful tool to solve problems in areas such as production, operation, finances, distribution of transport, among others. A linear programming model consists of the following:

Variables   These are the unknown variables that must be determined to solve a problem.

Restrictions   They are expressed as equations or inequalities, they are always linear functions and they are the limitations of the model.

Objective Function   It is expressed as a linear function that defines the effectiveness of the system, based on the variables and decision-making parameters.

Parameters   They are the known values in the model that link the variable with the restrictions and the objective function.

The graphic method is one of the procedures for solving problems of linear programming, it consists of graphing the restrictions of the problem to find the feasible region F_O. This is defined as the intersection of the regions that individually meet each one of the restrictions. F_O is a convex set since the subspaces formed by the restrictions are convex and a group of convex spaces is convex. Once F_O is determined, the problem of finding the optimum value is reduced to check the values of the vertices. Since it is a convex region, the optimum solution must be in one of the vertices. Using matrix notation, a linear programming problem in canonical form is expressed as follows:

Note 5.1. A set is convex if given any two points of the set, the segment joining them is included in this set.

$$max \ F_O(x,y) = c_i x + c_j \text{restrictedto:} Ax_i \leq b, x_i \geq 0 \qquad \textbf{(5.2.)}$$

Where $c_i \in R^n$ is the cost or profit vector, $x_i \in R^n$ is the decision variables vector, A is the restriction matrix ($m \times n$), and $b_i \in R^m$ is the vector of independent terms or resources ($n \times 1$). A graph is usually used to assess the solution space.

Example 5.3. A retail store asks a manufacturer some sports jackets and pants. The manufacturer has 750m of cotton fabric and 1000m of polyester fabric for their production. Each pair of pants needs 1m of cotton and 2m of polyester. Each jacket needs 1.5m of cotton and 1m of polyester. The cost of each pair of pants is $50 and the cost of a jacket is $40. (i) How many pants and jackets the manufacturer needs to supply to get maximum profit? (ii) How many to get minimum profit? (iii) Is $F_O(x,y) = c_1 x + c_2 y, c_i \in \mathbb{R}$ a linear transformation?

Solution 5.3. (i) If x = the number of pants and y = the number of jackets.

$$F_O(x,y) = 50x + 40y$$

Restricciones:

$$x + 1.5y \leq 750$$

$$2x + y \quad \leq 1000$$

$$x \quad \geq 0$$

$$y \quad \geq 0$$

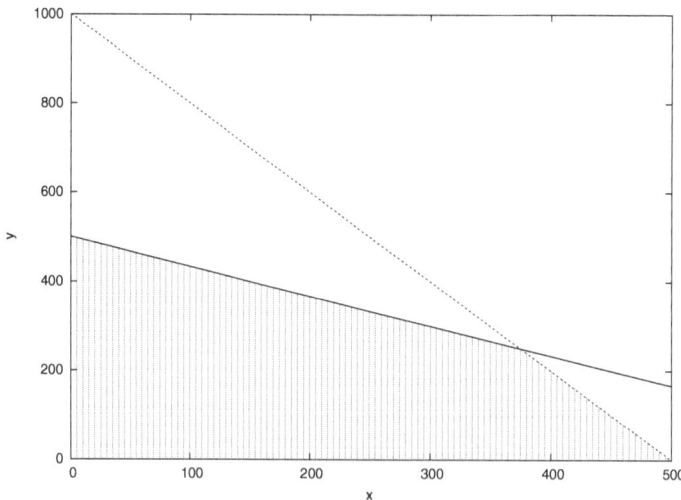

Fig. (5.1). Plot of functions $f_1(x) = \dfrac{1500}{3} - \dfrac{2}{3}x$ (continuous line), $f_2(x) = 1000 - 2x$. (dotted line), F_O (shaded region).

The shaded area is F_0, thus, the problem is reduced to verify the values of the vertices in the objective function to find the maximum $(x_1, y_1) = (0,500) \Rightarrow z_1 = 50x_1 + 40y_1 = 20,000$. $(x_2, y_2) = (375,250)$ $z_2 = 50x_2 + 40y_2 = 25,000$. $(x_3, y_3) = (500,0) \Rightarrow z_3 = 50x_3 + 40y_3 = 28,750$. Therefore, the optimum solution would be (x_2, y_2), producing 375 pants and 250 jackets, with maximum profit of \$28,750. (ii) Evaluating the vertices of the objective function we find that $(x_1, y_1) = (0,500)$ the minimum profit of \$20,000. can be obtained (iii) $\alpha F_0(x, y) = \alpha c_1 x + \alpha c_2 y = F_0(\alpha x, \alpha y)$; $F_0(x_1, x_2) + F_0(x_2, y_2) = c_1 x_1 + c_1 x_2 + c_2 y_1 + c_2 y_2 = c_1(x_1 + x_2) + c_2(y_1 + y_2) = F_0(x_1 + x_2, y_1 + y_2)$.   So $F_0(x, y)$ is a linear transformation.

CHAPTER 6

Bioinformatics

Carlos Polanco[*]

Faculty of Sciences, Universidad Nacional Autónoma de México, México

Abstract: Bioinformatics is a scientific field that combines mathematics and computer sciences to generate algorithms that model biological and medical scenarios. This chapter describes the basic properties of algorithms, their main classification as well as their restrictions and extensions.

Keywords: Bioinformatics, genome, nonlinear transformations, nucleotides, polar profile, proteins, structural proteomics.

6.1. BACKGROUND

This scientific discipline was born with the sequencing of the human genome and the knowledge of the protein structure *i.e.*, linear primary, secondary, tertiary, and quaternary representations [Polanco, 2013]. The analysis of these structures is done at two levels: amino acids and nucleotides. Both these levels are interconnected by different equivalence rules, the first where a triplet of nucleotides is equivalent to an amino acid and the second which is the set of 20 amino acids nature uses to form a protein.

6.2. STRUCTURAL PROTEOMICS

In the last three decades, two types of methods have been developed: one using a linear epresentation and the other with the tertiary representation of proteins. Both of them are equally efficient. The method here introduced, named Polarity Index Method [Polanco, 2014, 2016a,b,b; Polanco *et al.*, 2013a,b, 2014a, 2012, 2016a, 2014b,c,d,e, 2013c, 2015b, 2016b, 2014f, 2016c], is based on the linear epresentation and it only uses one physico-chemical property, the polarity [Pauling, 1995]. The Polarity Index Method is a supervised method, its metrics calculate the polar profile of the representative protein group and the target protein. The polar profile is calculated replacing each amino acid in the protein with a numeric value

[*]**Corresponding author Carlos Polanco:** Faculty of Sciences, Universidad Nacional Autónoma de México, México City, México; Tel: +01 55 5622 4858; Fax: +01 5556 4859; E-mail: polanco@unam.mx

according to the rule {P, P-, N, NP} = {1,2,3,4}. The new numeric representation is read from left to right, in pairs, one amino acid at a time. The incidences are represented in an incidence matrix A of the training protein group. The polar profile of the target protein B, which is the protein studied, is calculated the same way. Finally matrix B is weighted with the representative protein group A, adding them up $A + B$. If the difference between matrix A and $A + B$ is less than 25%, the method considers the target protein associated with the representative protein group. The metrics of this method represent a nonlinear transformation due to the unequal weighted of matrix $a_{i,j} + b_{i,j}$.

Example 6.1.   Be protein MQNVINTVKGKALEVAEYLT, its numeric representation is 33443341314424423434. This protein is one in a protein group of 120 proteins.

$$A = \begin{pmatrix} 17 & 43 & 21 & 65 \\ 5 & 3 & 8 & 4 \\ 0 & 1 & 0 & 7 \\ 11 & 8 & 23 & 5 \end{pmatrix} \quad B = \begin{pmatrix} 0 & 0 & 1 & 1 \\ 0 & 0 & 1 & 4 \\ 1 & 0 & 2 & 4 \\ 1 & 1 & 2 & 2 \end{pmatrix} \Rightarrow T(B) = A + B$$

$$= \begin{pmatrix} 17 & 43 & 22 & 66 \\ 5 & 3 & 9 & 8 \\ 1 & 1 & 2 & 9 \\ 12 & 9 & 25 & 7 \end{pmatrix}$$

In matrix B, element (2,4) located in row 2 has the maximum value, but in matrix $A + B$ the maximum value in this row corresponds to element (2,3).

6.3. GENETICS OF DISEASE

A protein is considered a functional unit according to its specialization or its pathogenic action. However, this pathogenic action is not unique but preferential [Wang and Wang, 2009] *i.e.*, a protein can show toxicity towards bacteria and also towards virus [Horowitz, 1996]; and although it can have action on several pathogen agents, a preference always stands out. This preference is measured by the amount of that protein required to prevent the growth of any pathogen agent, *e.g.*, cancer cells, fungi, or bacteria [Polanco, 2014, 2016a,b,b; Polanco *et al.*, 2013a,b, 2014a, 2012, 2016a, 2014b,c,d,e, 2013c, 2015b, 2016b, 2014f, 2016c].

This leads us to the assumption that in the protein structure (in its linear structure or in its tertiary structure), exists small differences in the order of the amino acids or in the type of amino acids that modify this specialization.

Example 6.2. Find the coincidences in the 20 peptides (short proteins) associated to cancer cells [Polanco *et al.*, 2015a] from APD2 Database [Wang and Wang, 2009]. Is this profile nonlinear?

Solution 6.1. From rule {P, P-, N, NP} = {1,2,3,4} the 20 peptides Table 6.1 were converted to their numeric equivalence and the relative frequency distribution was calculated (see Section 6.2), Fig. (**6.1**) shows a tendency in the maximum and minimum points located.

Table 6.1. Peptides list

Sequence	Numeric format
GEFLKCGESCVQGECYTPGCSCDWPICKKN	324413323343323234333324443113
GEYCGESCYLIPCFTPGCYCVSRQCVNKN	322332332444343433234313343313
GIACGESCVFLGCFIPGCSCKSKVCYFN	344332334443344433313143243
GIPCAESCVWIPCTVTALIGCGCSNKVCYN	344342334444334344433333314323
GIPCAESCVWIPPCTITALMGCSCKNNVCYNN	344342334444334344433331334 3233
GLLPCAESCVYIPCLTTVIGCSCKSKVCYKN	344434233424434334443333 13143213
GLPVCGETCAGGTCNTPGCSCSWPICTRN	344433233433333343333344443313
GLPVCGETCFGGTCNTPGCTCDPWPVCTRN	344433233433333343333244443313
GLPVCGETCVGGTCNTPGCSCSWPVCTRN	344433233433333343333344443313
GTFPCGESCVFIPCLTSAIGCSCKSKVCYKN	334433233444434334443333 13143213
GVIPCGESCVFIPCISSVLGCSCKNKVCYRD	344433233444434334443333 13143212
GVPICGETCTLGTCYTAGCSCSWPVCTRN	344433233343332343333344443313
KLAKLAKKLAKLAK	14414411441441
KLCGETCFKFKCYTPGCSCSYPFCK	143323341413234333332443 1
KSCCKNTTGRNIYNTCRFAGGSRERCAKLSGCK IISASTCPSDYPK	133313333134233314433312134143331443433343 2 241

(Table 6.1) cont.....

KSCCPNTTGRNIYNTCRFAGGSRERCAKLSGCKIISASTCPSDYPK	1333433331342333144333121341433314434333432241
KSCCPNTTGRNIYNTCRFGGGSREVCARISGCKIISASTCPSDYPK	1333433331342333143333124341433314434333432241
KSCCPNTTGRNIYNTCRFGGGSRQVCASLSGCKIISASTCPSDYPK	1333433331342333143333134343433314434333432241
KSCCPNTTGRNIYNTCRLGGGSRERCASLSGCKIISASTCPSDYPK	1333433331342333143333121343433314434333432241
KSCCPNTTGRNIYNTCRLTGSSRETCAKLSGCKIISASTCPSNYPK	1333433331342333143333123341433314434333433241

ᵃ Linear sequence peptides associated to cancer cells. Source [Polanco *et al.*, 2015a; Wang and Wang, 2009].

The polar profile in Fig. (**6.1**) determines a bias for a particular protein group, *i.e.*, if there were no bias, the 16 polar interactions would have the same frequency. Is it possible that different protein groups have the same polar profile? No, it is not. Proteins with similar main function have similar polar profile [Polanco, 2014, 2016a,b,b; Polanco *et al.*, 2013a,b, 2014a, 2012, 2016a, 2014b,c,d,e, 2013c, 2015b, 2016b, 2014f, 2016c]. If amino acids are added to the protein, the polar profile does not necessarily change *i.e.*, $T(x + y) \neq T(x) + T(y)$. So this transformation is nonlinear.

6.4. MARKOV MODEL

This method uses the Markov conjeture [Eddy, 1996a] to evaluate several variables simultaneously over the same linear sequence. Therefore, it does not need to know all the information of the variables involved, only its present value. This stochastic method adds to the solution a probability value. Particularly, the Hidden Markov model is the most efficient, although it is more complex to implement as it requires to identify the "cause" of the variables involved.

Example 6.3. Apply Hidden Markov model to identify the Selective Cationic Amphiphatic Antibacterial Peptides profile. Is this profile nonlinear?

Solution 6.2. Selective Cationic Amphipathic Antibacterial Peptides (SCAAP) [Polanco and Samaniego-Mendoza, 2009] are short natural proteins (Table **6.2**) with strong pathogenic action over bacteria membranes and weak pathogenic action against mammalian cells. SCAAP group is very important in drug design, particularly in the production of antibiotics. The identification of this group implied to find the relation between phisical-chemical parameters [Polanco and Samaniego-Mendoza, 2009] and the SCAAP polar profile [Polanco *et al.*, 2013a].

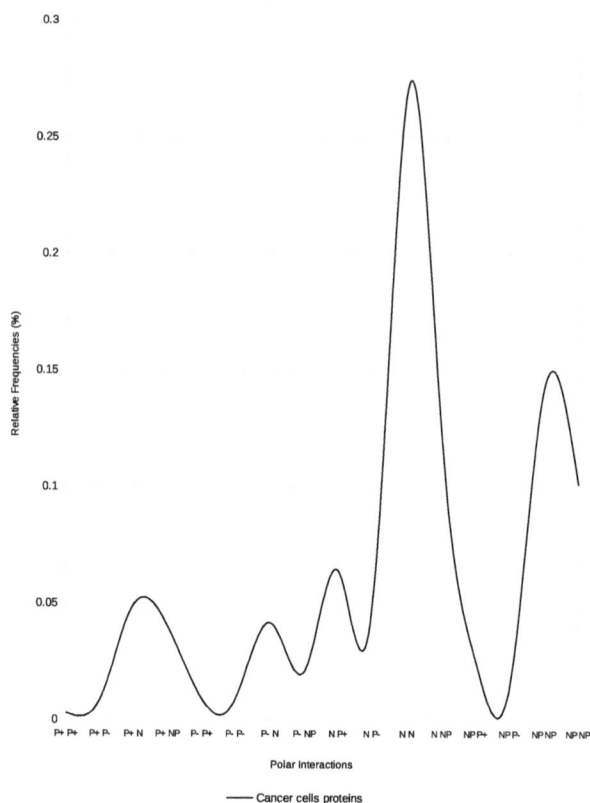

Fig. (6.1). Relative frecuency distribution peptides see Table **6.1**.

Table 6.2. SCAAP peptides list.

Sequence	Linear sequence
1 KIAKKIAKIAKKIA-NH2	KIAKKIAKIAKKIA
(KIAKKIA)3NH2 KIAKKIAKIAKKIAKIAKKIA-NH2	KIAKKIAKIAKKIAKIAKKIA
(KIAKLAK)2NH2 KIAKLAKKIAKLAK-NH2	KIAKLAKKIAKLAK
(KIAKLAK)3NH2 KIAKLAKKIAKLAKKIAKLAK-NH2	KIAKLAKKIAKLAKKIAKLAK
(KALKALK)3NH2 KALKALKKALKALKKALKALK-NH2	KALKALKKALKALKKALKALK
(KLGKKLG)3NH2 KLGKKLGKLGKKLGKLGKKLG-NH2	KLGKKLGKLGKKLGKLGKKLG

(Table 6.2) cont.....

CecropinA KWKLFKKIEKVGQNIRDGIIKAGPAVAVVGQAT QIAK-NH2	KWKLFKKIEKVGQNIRDGIIKAGPAVAVVGQAT QIAK
Melittin GIGAVLKVLTTGLPALISWIKRKRQQ-NH2	GIGAVLKVLTTGLPALISWIKRKRQQ
Magainin 2 GIGKFLHSAKKFGKAFVGEIMNS-NH2	GIGKFLHSAKKFGKAFVGEIMNS
CA(1â€"13)M(1â€"13)NH2 KWKLFKKIEKVGQGIGAVLKVLTTGL-NH2	KWKLFKKIEKVGQGIGAVLKVLTTGL
CA(1â€"8)M(1â€"18)NH2 KWKLFKKIGIGAVLKVLTTGLPALIS-NH2	KWKLFKKIGIGAVLKVLTTGLPALIS
Kla1 KLALKLALKAWKAALKLA-NH2	KLALKLALKAWKAALKLA
Kla2 KLALKAALKAWKAAAKLA-NH2	KLALKAALKAWKAAAKLA
Kla3 KLALKAAAKAWKAAAKAA-NH2	KLALKAAAKAWKAAAKAA
Kla7 KAIAKSILKWIKSIAKAI-NH2	KAIAKSILKWIKSIAKAI
Kla8 KALAALLKKWAKLLAALK-NH2	KALAALLKKWAKLLAALK
Kla9 KLLAKAALKWLLKALKAA-NH2	KLLAKAALKWLLKALKAA
Kla10 KALKKLLAKWLAAAKALL-NH2	KALKKLLAKWLAAAKALL
Kla11 KITLKLAIKAWKLALKAA-NH2	KITLKLAIKAWKLALKAA
Kla12 KALAKALAKLWKALAKAA-NH2	KALAKALAKLWKALAKAA
m2a GIGKFLHSAKKFGKAFVGEIMNS-NH2	GIGKFLHSAKKFGKAFVGEIMNS
W16-m2a GIGKFLHSAKKFGKAWVGEIMNS-NH2	GIGKFLHSAKKFGKAWVGEIMNS
L2R11A20-m2a GLGKFLHSAKRFGKAFVGEAMNS-NH2	GLGKFLHSAKRFGKAFVGEAMNS
I6L15-m2a GIGKFIHSAKKFGKLFVGEIMNS-NH2	GIGKFIHSAKKFGKLFVGEIMNS
I6A8L15I17-m2a GIGKFIHAAKKFGKLFIGEIMNS-NH2	GIGKFIHAAKKFGKLFIGEIMNS
I6R11R14W16-m2a GIGKFIHSAKRFGRAWVGEIMNS-NH2	GIGKFIHSAKRFGRAWVGEIMNS
I6V9W12T15I17-m2a GIGKFIHSVKKWGKTFIGEIMNS-NH2	GIGKFIHSVKKWGKTFIGEIMNS
-m2a GIAKFGKAAAHFGKKWVGELMNS-NH2	GIAKFGKAAAHFGKKWVGELMNS
-m2a GIGKFLHTLKTFGKKWVGEIMNS-NH2	GIGKFLHTLKTFGKKWVGEIMNS
-m2a GIGHFLHKVKSFGKSWIGEIMNS-NH2	GIGHFLHKVKSFGKSWIGEIMNS

[a] Linear sequence SCAAP peptides. Source [Polanco and Samaniego-Mendoza, 2009].

To find the solution we follow the next procedure: Step 1. Calculate the Isoelectric point, Helical hydrophobic moment, Mean hydrophobicity and Mean net charge from its linear protein sequence [Polanco *et al.*, 2012]. Calculate the polar profile of the SCAAP group [Polanco *et al.*, 2013a]. Compare the sequences accepted by both procedures. Because the number of hits is equal, both procedures are equivalent. Step 1 uses a numerical range to accept or reject the protein, thus, $\Rightarrow$ $T(x + y) \neq T(x) + T(y)$. Step 2 the polar profile not necessary change when amino acids are added to the protein *i.e.*, $T(x + y) \neq T(x) + T(y)$. Therefore, both methods represent a nonlinear transformation.

CHAPTER 7

Calculus

Carlos Polanco[*]

Faculty of Sciences, Universidad Nacional Autónoma de México, México

Abstract: An integral, as a mathematical operator, is substantially simplified by transformations over the integration area. This chapter defines the main types of transformations, surface integrals, Â line integrals, and the Jacobian determinant.

Keywords: Composition transformations, Gauss theorem, Green theorem, Jacobian determinant, Curved integrals, Matrix derivatives, Reflection transformations, Rotation transformations, $1D$ Transformations, $1D$ to $2D$ Transformations, $2D$ Transformations, $2D$ to $3D$ Transformations, $3D$ Transformations, Stokes theorem, Surface integrals, Vector calculus.

7.1. BACKGROUND

Reflection and Rotation transformations play an important role in the Stokes, Gauss, and Green theorems [Marsden and Tromba, 2011] Ch. 8, simplifying the integrand and integration area in *curved integrals*, and *surface integrals*.

7.2. JACOBIAN DETERMINANT

If $f: \mathbb{R}^m \to \mathbb{R}^n$, and T is a transformation such that $T: \mathbb{R}^n \to \mathbb{R}^n$, the Jacobian determinant or Jacobian $J(T)$ (7.1) is the *matrix derivatives* of T.

$$J(T) = J_T(x_1, x_2, \cdots, x_n) = \frac{\partial(T_i, T_2, \cdots, T_n)}{\partial(x_1, x_2, \cdots, x_n)} = \begin{vmatrix} \frac{\partial T_1}{\partial x_1} & \frac{\partial T_1}{\partial x_2} & \cdots & \frac{\partial T_1}{\partial x_n} \\ \frac{\partial T_2}{\partial x_1} & \frac{\partial T_2}{\partial x_2} & \cdots & \frac{\partial T_2}{\partial x_n} \\ \vdots & \vdots & \ddots & \vdots \\ \frac{\partial T_n}{\partial x_1} & \frac{\partial T_n}{\partial x_2} & \cdots & \frac{\partial T_n}{\partial x_n} \end{vmatrix} \qquad (7.1)$$

[*]**Corresponding author Carlos Polanco:** Faculty of Sciences, Universidad Nacional Autónoma de México, México City, México; Tel: +01 55 5622 4858; Fax: +01 5556 4859; E-mail: polanco@unam.mx

Example 7.1. Be $T(x,y) = (2x, 3y)$, $x \in [0,1]$, $y \in [0,1]$. (i) What is $J(T)$? (ii) Is T linear?

Solution 7.1. (i) $J(T) = \begin{vmatrix} 2 & 0 \\ 0 & 3 \end{vmatrix} = 6$ (ii) $T(x_1 + x_2, y_1 + y_2) = (2x_1 + 2x_2, 3y_1 + 3y_2) = T(x_2, y_1) + T(x_2, y_2)$, $\alpha T(x,y) = (\alpha 2x, \alpha 3y) = T(\alpha x, \alpha y)$. (ii) T is a linear transformation.

Example 7.2. Be $T_1(x,y) = (x,y)$. (i) What is $J(T_1)$? (ii) If $T_2(r,\theta) = (r\cos\theta, r\sin\theta)$, where $r \in [0,1]$, $\theta \in [0,2\pi]$ what is $J(T_2)$?

Solution 7.2. (i) $J(T_1) = \begin{vmatrix} 1 & 0 \\ 0 & 1 \end{vmatrix} = 1$ (ii) $J(T_2) = \begin{vmatrix} \cos\theta & -r\sin\theta \\ \sin\theta & r\cos\theta \end{vmatrix} = r\cos^2\theta + r\sin^2\theta = r$.

7.3. 1D TRANSFORMATION

A T transformation from $\mathbb{R}$ to $\mathbb{R}$ converts subsets from $\mathbb{R}$ to $\mathbb{R}$. T is an endomorphism. See section 4 to calculate the area on the plane trajectories.

Note 7.1. This transformation simplifies the integrand of the integral, *i.e.*, $\int_a^b f \circ T |J(T)| \, dt = \int_c^d f(x) \, dx$,   where $T: [t_0, t_1] = [a,b] \subset \mathbb{R} \to [x_0, x_1] = [T(t_0), T(t_1)] = [c,d] \subset \mathbb{R}$.

Example 7.3. Be $\int_2^4 xe^{x^2} \, dx$. (i) Use this transformation $T: [4,16] \subset \mathbb{R} \to [2,4] \subset \mathbb{R}: \sqrt{t}$ to solve the integral. (ii) Is T linear? (iii) Use the integration by substitution method to solve the integral.

Solution 7.3. (i) $T(t) = \sqrt{t}$, $t \in [4,16] \Rightarrow f(x) \circ T(t) = f(T(t)) = f(\sqrt{t}) = \sqrt{t}e^t$. $|J(T)| = |-\frac{1}{2\sqrt{t}}| = \frac{1}{2\sqrt{t}}$, then $\int_{x_0}^{x_1} xe^{x^2} \, dx = \int_2^4 xe^{x^2} \, dx = \int_4^{16} \sqrt{t}e^t \frac{1}{2\sqrt{t}} \, dt = \frac{1}{2}\int_4^{16} e^t \, dt = \frac{1}{2}(e^{16} - e^4)$. (ii) $T(x_1 + x_2) = \sqrt{x_1 + x_2} \neq \sqrt{x_1} + \sqrt{x_2}$. So T is nonlinear. (iii) If $u = x^2 \Rightarrow du = 2x \, dx$, then $\int_2^4 xe^{x^2} \, dx; u_0 = 4, u_1 = 16$, which is equivalent to $\int_4^{16} xe^u \frac{du}{2x} = \frac{1}{2}\int_4^{16} e^u \, du = \frac{1}{2}(e^{16} - e^4)$.

7.4. 1D To 2D TRANSFORMATION

This transformation has an important application over a curved integral. Be $T: [t_0, t_1] \subset \mathbb{R} \to \mathbb{R}^2$, the curved integral is defined by $\oint_{t_0} F \circ T(t) \cdot T'(t) \, dt$. The physical meaning of a curve integral is the work done to travel a T path, from point t_0 to point t_1, under the influence of a vector field F.

Note 7.2. Integral along a curve in space. Geometrically ?, when a path is defined over a plane *i.e.*, $T: \mathbb{R} \to \mathbb{R}^2$, its graph f is a surface in 3D space. The *integral along a curve in space* gives the cross-sectional area bounded by the curve $T: \mathbb{R} \to \mathbb{R}^2$ and the graph of $f \circ T$, *i.e.*, $\int_{t_0}^{t_1} f \circ T(t) \, \|T'(t)\| \, dt$.

Example 7.4. Suppose [Farmer *et al.*, 2012] Ex. 16.2.3, an object moves from (â|四 to (2,4) along the path $r(t) = (t, t^2)$, subject to force $F = (x \sin y, y)$. (i) Find the work done. (ii) Is $r(t)$ linear? (i) We can write the force $F: \mathbb{R}^2 \to \mathbb{R}^2, (x \sin y, y)$ and the path $r: \mathbb{R} \to \mathbb{R}^2, (t, t^2)$, in terms of t as $F \circ r = F(r(t)) = F(t, t^2) = (t \sin t^2, t^2)$ and compute $r'(t) = (1, 2t)$. If $(t, t^2) = (1, 1) \Rightarrow t_0 = -1$; if $(t, t^2) = (2,4) \Rightarrow t_1 = 2$. Then the work is done by $\oint_{t_0} F \circ r(t) \cdot r'(t) \, dt = \oint_{-1} (t \sin t^2, t^2) \cdot (1, 2t) \, dt = \oint_{-1} (t \sin t^2 + 2t^3 \, dt = \frac{15}{2} + \frac{\cos(1)}{2} - \frac{\cos(4)}{2} = 8.09$. t_0 and t_1 are given in radians. (ii) $r(x_1 + x_2) = (x_1 + x_2, (x_1 + x_2)^2) = (x_1 + x_2, x_1^2 + x_2^2 + 2x_1 x_2) \neq r(x_1) + r(x_2) = (x_1, x_1^2) + (x_2, x_2^2) = (x_1 + x_2, x_2^2 + x_1^2)$. So r is nonlinear.

Example 7.5. Be transformation $T: [0, \frac{\pi}{2}] \subset \mathbb{R} \to \mathbb{R}^2, (t, t)$ and function $f: \mathbb{R}^2 \to \mathbb{R}, x^2 y^3$ (note that the graph is in 3D space, and transformation T is in 2D space). (i) Determine the curve integral along a curve in space. (ii) Is T linear?

Solution 7.5. (i) According to the path and the function described $f \circ T = f(t) = f(t, t) = t^2 t^3$, $\quad T'(t) = (1, 1); \quad \int_0^{\frac{\pi}{2}} f \circ T(t) \|T'(t)\| dt = \int_0^{\frac{\pi}{2}} t^5 \sqrt{2} dt = \sqrt{2} \int_0^{\frac{\pi}{2}} t^5 \, dt = \sqrt{2} \left[\frac{t^6}{6}\right]_0^{\frac{\pi}{2}} = 2.5$. (ii) $T(x_1 + x_2) = (x_1 + x_2, x_1 + x_2) = (x_1, x_1) + (x_2, x_2) = T(x_1) + T(x_2)$, $\alpha T(x_1) = \alpha(x_1, x_1) = (\alpha x_1, \alpha x_1) = T(\alpha x_1)$. (ii) T is linear.

Example 7.6. Be transformation $T: [0, \frac{\pi}{2}] \hat{A} \subset \mathbb{R} \to \mathbb{R}^2, (4 \cos t, 4 \sin t)$ and function $f: \mathbb{R}^2 \to \mathbb{R}, xy^3$ (note the graph is in 3D and transformation T is in 2D). (i) Determine the integral along a curve in space.(ii) Is T linear?

Solution 7.6. (i) Be the composition transformation $f \circ T = f(4\cos t, 4\sin t) = 16\cos^2(t)\sin^3 t,$ $T'(t) = (-4\sin t, 4\cos t);$ $||T'(t)|| = 4$ $\Rightarrow$ $\int_0^{\frac{\pi}{2}} f \circ T(t)Â||T^{Â'}(t)||dt$ $= 4\int_0^{\frac{\pi}{2}} \cos t \sin^3 t \, dt = 4\int_0^1 u^3 \, du = 4[\frac{u^4}{4}]_0^1 = 1.$ (ii) T is nonlinear because $\sin x$ and $\cos x$ are nonlinear functions.

7.5. 2D TRANSFORMATION

A transformation T from $\mathbb{R}^2$ to $\mathbb{R}^2$ is an endomorphism of a plane. This very common transformation, transforms the y/x plane to the x/y plane. It requires two transformations to transform the y/x coordinate plane (where the y axis is the vertical line), into the x/y coordinate plane (where the x axis is the vertical line). First, a reflection transformation will be used (y–axis π radians) $T_{re}(x, y) = (-x, y)$ then, a rotation transformation will be used (clockwise rotation by $\pi/2$) $T_{ro}(x, y) = (y, -x)$. So $T_{x/y}(x, y) = T_{ro} \circ T_{re} = T_{ro}(T_{re})(x, y) = T_{ro}(-x, y) = (y, x)$.

Note 7.3. The S^* area in the y/x coordinate plane is converted to the S area in the x/y coordinate plane through the reflection and rotation transformations, *i.e.*, $S = T_{x/y}(S^*)$, $\forall(x, y) \in S^*$ area. These transformations preserve the area and shape of S^*.

Example 7.7. Be $\iint_S dy dx$, where $S = \{(x, y) | x \in [-1, 1], y \in [-2, 2]\}$ is equivalent to the integral $\iint_{S^*} dx \, dy$, and $S^* = T_{x/y}(S)$, $\forall(x, y) \in S$. (i) What is S^* area? (ii) Calculate the integrals $\iint_{S^*} dy \, dx$ and $\iint_{S^*} dxdy$ (iii) Is $T(x, y)$ a linear transformation?

Solution 7.7. (i) Because the integration area is a paralelogram, it is possible to determine the new paralelogram with $T_{x/y}(x, y)$. So $T_{x/y}(-1, 2) = (2, -1)$, $T_{x/y}(1, 2) = (2, 1)$, $T_{x/y}(-1, -2) = (-2, -1)$ and $T_{x/y}(1, -2) = (-2, 1)$. (ii) $\int_{x_0}^{x_1} \int_{y_0(x)}^{y_1(x)} dy \, dx = \int_{-1}^1 \int_{-2}^2 dy \, dx = \int_{-1}^1 [\int_{-2}^2 dy] \, dx = 4\int_{-1}^1 dx = 4(2) = 8.$ $\int_{y_0}^{y_1} \int_{x_0(y)}^{x_1(y)} dx \, dy = \int_{-2}^2 \int_{-1}^1 dx \, dy = \int_{-2}^2 [\int_{-1}^1 dx] \, dy = 2\int_{-2}^2 dy = 2(4) = 8$ (iii) $T_{x/y}(x_1 + x_2, y_1 + y_2) = (y_1 + y_2, x_1 + x_2) = T_{x/y}(x_1, y_1) + T_{x/y}(x_2, y_2),$ $\alpha T_{x/y}(x, y) = \alpha(y, x) = (\alpha x, \alpha y) = T_{x/y}(\alpha x, \alpha y)$. So $T_{x/y}(x, y)$ is linear.

Here, a transformation is introduced that neither preserves the area nor the shape. It is called Polar transformation *i.e.*, $T: [0, 1] \times [0, 2\pi] \subset \mathbb{R}^2 \to \mathbb{R}^2, (r\cos\theta, r\sin\theta)$.

Note 7.4. The S^* area in the r/θ coordinate plane is converted to the S area in the x/y coordinate plane through transformation T, *i.e.*, $S = T_{r/\theta}(S^*)$, $\forall (r,\theta) \in S^*$ area. This transformation does **not** preserve neither the area nor the shape of S^*. In this case it is necessary to use general format, $\iint_{S^*} f \circ T \, |J(T)| \, dr \, d\theta = \iint_S f(x,y) \, dx \, dy$.

Example 7.8. Be transformation $T: [a,b] \times [0,\frac{\pi}{2}] \subset \mathbb{R}^2 \to \mathbb{R}^2, (r\cos\theta, r\sin\theta)$, where $a < b$. Use this transformation to calculate the double integral $\int_0^1 \int_0^{\sqrt{1-x^2}} x^2 + y^2 dy \, dx$, with the general format.

Solution 7.8. Integral $\int_0^1 \int_0^{\sqrt{1-x^2}} x^2 + y^2 dy \, dx$ is equivalent to $\int_a^b \int_0^{\frac{\pi}{2}} r^2 \, |r| d\theta \, dr$
$= \frac{\pi}{2} \int_a^b r^3 \, dr = \frac{\pi}{2} \left[\frac{r^4}{4}\right]_a^b = \frac{\pi}{2} \left[\frac{(b^4-a^4)}{4}\right]$. Note that $J(T) = \begin{vmatrix} \cos\theta & -r\sin\theta \\ \sin\theta & r\cos\theta \end{vmatrix} = r\cos^2\theta + r\sin^2\theta = r$.

7.6. 2D TO 3D TRANSFORMATION

The T transformation here described, has an important application over a surface integral. Be $T: D \subset \mathbb{R}^2 \to S \subset \mathbb{R}^3$, the surface integral will be defined by $_DF \circ T \cdot T_u \times T_v \, du \, dv$, where $D = [u_0, u_1] \times [v_0, v_1]$. The surface integral measures the effect of an F vector field on an S surface.

Note 7.5. Oriented Surfaces "An **oriented surface** is a two-sided surface with one side specified as **outside** or **positive side**; and the other side called **inside** or **negative side**. At each point $(x,y,z) \in S$ there are two unit normal vectors n_1 and n_2, where $n_1 = n_2$. Each of these two normals can be associated with one side of the surface. Thus, to define a side of surface S, we choose for each point a unit normal vector n pointing the positive side of surface S ? Ch. 7.6, p. 402.

Example 7.9. Let D be a rectangle in the θ/ϕ plane defined by $0 \leq \theta \leq 2\pi$, $0 \leq \phi \leq \pi$ and let surface S be defined by transformation T_{Sp} (Thus S is the unit sphere parametrized by T). Let $F(x,y,z) = (x,y,z)$? Ch. 7.6. Compute $_DF \circ T \cdot T_\theta \times T_\phi \, d\theta \, d\phi$.

Solution 7.9. Be $T_{Sp}: \mathbb{R}^3 \to \mathbb{R}^3$, $(\rho\sin\phi\cos\theta, \rho\sin\phi\sin\theta, \rho\cos\phi)$ and vector field $F(x,y,z) = (x,y,z)$; thus $F \circ T_{Sp} = T_{Sp}$, $T_\theta = \frac{\partial T_{Sp}}{\partial \theta}$, $T_\phi = \frac{\partial T_{Sp}}{\partial \phi} \Rightarrow T_\theta \times T_\phi = (-\sin^2\phi\cos\theta, \sin^2\phi\cos\theta, -\sin\phi\cos\phi)$. The surface integral is $_D F \circ T \cdot T_\theta \times T_\phi \, d\phi \, d\theta$, where $D = [\phi_0, \phi_1] = [0, \pi] \times [\theta_0, \theta_1] = [0, 2\pi]$; $\int_0^{2\pi} \int_0^{\pi} - \sin\phi \, d\phi \, d\theta = -2 \int_0^{2\pi} d\theta = -4\pi$.

7.7. 3D TRANSFORMATION

A T transformation from $\mathbb{R}^3$ to $\mathbb{R}^3$ is an endomorphism. Here are introduced two T transformations from $\mathbb{R}^3$ to $\mathbb{R}^3$ that preserve neither the area nor the shape: Cylindrical transformation $T_{Cy}: \mathbb{R}^3 \to \mathbb{R}^3$, $(r\cos\theta, r\sin\theta, z)$, and Spherical transformation $T_{Sp}: \mathbb{R}^3 \to \mathbb{R}^3$, $(\rho\sin\phi\cos\theta, \rho\sin\phi\sin\theta, \rho\cos\phi)$. A 3D space is a relation $\mathbb{R} \times \mathbb{R} \times \mathbb{R}$, where each axis is represented by a real axis, named x axis, y axis, and z axis, that make possible six different 3D spaces. Be $T: \mathbb{R} \to \mathbb{R}^3$, the real function $f: \mathbb{R}^3 \to \mathbb{R}$ and $J(T)$ the Jacobian determinant: (1) $z/y/x$ space is associated with the triple integral $\int_{x_0}^{x_1} \int_{y_0(x)}^{y_1(x)} \int_{z_0(x,y)}^{z_1(x,y)} f \circ T |J(T)| \, dz \, dy \, dx$, (2) $z/x/y$ space is associated with the triple integral $\int_{y_0}^{y_1} \int_{x_0(y)}^{x_1(y)} \int_{z_0(y,x)}^{z_1(y,x)} f \circ T |J(T)| \, dz \, dx \, dy$, (3) $y/x/z$ space is associated with the triple integral $\int_{z_0}^{z_1} \int_{x_0(z)}^{y_1(z)} \int_{y_0(z,x)}^{y_1(z,x)} f \circ T |J(T)| \, dy \, dx \, dz$, (4) $y/z/x$ space is associated with the triple integral $\int_{x_0}^{x_1} \int_{z_0(x)}^{z_1(x)} \int_{y_0(x,z)}^{y_1(x,z)} f \circ T |J(T)| \, dy \, dx \, dz$, (5) $x/y/z$ space is associated with the triple integral $\int_{z_0}^{z_1} \int_{y_0(z)}^{y_1(z)} \int_{x_0(z,y)}^{x_1(z,y)} f \circ T |J(T)| \, dx \, dy \, dz$, and (6) $x/z/y$ space is associated with the triple integral $\int_{z_0}^{z_1} \int_{z_0(y)}^{z_1(y)} \int_{x_0(y,z)}^{x_1(y,z)} f \circ T |J(T)| \, dx \, dz \, dy$.

Example 7.10. Solve triple integral $\iiint_S \sqrt{x^2 + y^2} \, dz \, dy \, dx$, where $S = (x,y,z) \subset \mathbb{R}^3$ such that $x \in [-1,1], y \in [-\sqrt{1-x^2}, \sqrt{1-x^2}], z \in [-2,2]$; and the T_{Cy} transformation. (i) Use $\int_{z_0}^{z_1} \int_{\theta_0(z)}^{\theta_1(z)} \int_{r_0(\theta,z)}^{r_1(\theta,z)} f \circ T_{Cy} |J(T_{Cy})| \, dr \, d\theta \, dz$. (ii) Is T_{Cy} linear? (iii) Calculate $J(T_{Cy})$.

Solution 7.10. (i) The triple integral $\int_{x_0}^{x_1} \int_{y_0(x)}^{y_1(x)} \int_{z_0(x,y)}^{z_1(x,y)} f(x,y,z) \, dz \, dy \, dx$ is converted to $\int_{z_0}^{z_1} \int_{\theta_0(z)}^{\theta_1(z)} \int_{r_0(\theta,z)}^{r_1(\theta,z)} f \circ T_{Cy} |J(T_{Cy})| \, dr \, d\theta \, dz$. Applying the integration limits we obtain $\int_{-2}^{2}\int_{0}^{2\pi}\int_{0}^{1} \sqrt{r^2\cos^2\theta + r^2\sin^2\theta} \, |J(T_{Cy})| \, dr \, d\theta \, dz =$

$\int_{-2}^{2}\int_{0}^{2\pi}\int_{0}^{1} r \, |r| \, dr \, d\theta \, dz = \int_{-2}^{2}\int_{0}^{2\pi}\int_{0}^{1} r^2 \, dr \, d\theta \, dz = 8\pi[\frac{r^3}{3}]_{0}^{1} = \frac{8\pi}{3}$ (ii) If T

transformation were linear it would mean that $T(r_1 + r_2, \theta_1 + \theta_2, z_1 + z_2) =$
$[(r_1 + r_2)\cos(\theta_1 + \theta_2), (r_1 + r_2)\sin(\theta_1 + \theta_2), z_1 + z_2] = [(r_1 +$
$r_2)(\cos\theta_1\cos\theta_2 - \sin\theta_1\sin\theta_2), (r_1 + r_2)(\sin\theta_1\cos\theta_2 - \sin\theta_1\cos\theta_2), z_1 + z_2]$
would be equal to $T(r_1, \theta_1, z_1) + T(r_2, \theta_2, z_2) = [r_1\cos\theta_1 + r_2\cos\theta_2, r_1\sin\theta_1 +$
$r_2\sin\theta_2, z_1 + z_2]$, but it is not. Therefore T is nonlinear, because: $(r_1 + r_2)\cos(\theta_1 +$

$\theta_2) \neq (r_1 + r_2)(\cos\theta_1\cos\theta_2 - \sin\theta_1\sin\theta_2)$ (iii) $J(T_{Cy}) = \begin{vmatrix} \cos\theta & -r\sin\theta & 1 \\ \sin\theta & r\cos\theta & 1 \\ 0 & 0 & 1 \end{vmatrix} =$

$r\cos^2\theta + r\sin^2\theta = r$.

Example 7.11. Solve triple integral $\iiint_{S} e^{(x^2+y^2+z^2)^{\frac{3}{2}}} dz \, dy \, dx$, where $S =$
$(x, y, z) \subset \mathbb{R}^3$ such that $x \in [-1,1]$, $y \in [-\sqrt{1-x^2}, \sqrt{1-x^2}]$, $z \in$
$[-\sqrt{1-x^2-y^2}, \sqrt{1-x^2-y^2}]$ [Marsden and Tromba, 2011] Ch. 6.2; and the
T_{Sp} transformation. (i) Use $\int_{\rho_0}^{\rho_1}\int_{\theta_0(\phi)}^{\theta_1(\phi)}\int_{\rho_0(\theta,\phi)}^{\rho_1(\theta,\phi)} f \circ T_{Sp} \, |J(T_{Sp})| \, d\rho \, d\theta \, d\phi$. (ii) Is
T_{Sp} linear? (iii) Calculate $J(T_{Sp})$.

Solution 7.11. (i) The triple integral $\int_{x_0}^{x_1}\int_{y_0(x)}^{y_1(x)}\int_{z_0(x,y)}^{z_1(x,y)} f(x, yz) \, dz \, dy \, dx$ is

converted to $\int_{\phi_1}^{\phi_1}\int_{\theta_0(\phi)}^{\theta_1(\phi)}\int_{\rho(\theta,\phi)}^{\rho_1(\theta,\phi)} f \circ T|J(T)| \, d\rho \, d\theta \, d\phi = \int_{\phi_1}^{\phi_1}\int_{\theta_0(\phi)}^{\theta_1(\phi)}\int_{\rho(\theta,\phi)}^{\rho_1(\theta,\phi)}$

$e^{\rho^3}|\rho^2\sin\phi| \, d\rho \, d\theta \qquad d\phi = \int_{0}^{1}\int_{0}^{\pi}\int_{0}^{\pi} \rho^2 e^{\rho^3}\sin\phi \, d\rho \, d\theta \qquad d\phi =$

$2\pi\int_{0}^{\pi}\int_{0}^{1} \rho^2 e^{\rho^3}\sin\phi \, d\rho \, d\phi = -2\pi\int_{0}^{1} \rho^2 \, e^{\rho^3}[\cos\phi]_{0}^{\pi} \, d\rho = 4\pi\int_{0}^{1} \rho^2 e^{\rho^3} \, d\rho =$

$\frac{4\pi}{3}\int_{0}^{1} e^u \, du = \frac{4\pi}{3}(e - 1)$. (ii) $T(\rho_1 + \rho_2, \theta_1 + \theta_2, \phi_1 + \phi_2) = T(\rho_1, \theta_1, \phi_1) +$
$T(\rho_2, \theta_2, \phi_2)$. Both should be equal, but they are not. (ii) So T is nonlinear because
$(r_1 + r_2)\cos(\theta_1 + \theta_2) \neq (r_1 + r_2)(\cos\theta_1\cos\theta_2 - \sin\theta_1\sin\theta_2)$.

(iii)

$$\begin{vmatrix} \sin\phi\cos\theta & -\rho\sin\phi\sin\theta & \rho\cos\phi\cos\theta \\ \sin\phi\sin\theta & -\rho\sin\phi\cos\theta & \rho\cos\phi\sin\theta \\ \cos\phi & 0 & -\rho\sin\phi \end{vmatrix} =$$

$$\cos\phi \begin{vmatrix} -\rho\sin\phi\sin\theta & \rho\cos\phi\cos\theta \\ -\rho\sin\phi\cos\theta & \rho\cos\phi\sin\theta \end{vmatrix} -$$

$$\rho\sin\phi \begin{vmatrix} \sin\phi\cos\theta & -\rho\sin\phi\sin\theta \\ \sin\phi\sin\theta & -\rho\sin\phi\cos\theta \end{vmatrix} = -\rho^2\sin\phi$$

CHAPTER 8

Cybernetics

Carlos Polanco[*]

Faculty of Sciences, Universidad Nacional Autónoma de México, México

Abstract: Cybernetics is especially focused on processing since our telecommunications and computers work with electromagnetic signals processed through electronic circuits that produce specific responses under specific conditions that can encrypt information. This chapter reviews three issues: signal processing, artificial neural network, and natural language processing.

Keywords: Artificial neural network, Laplace transforms, Linear transformation, Natural language processing, Nonlinear transformation, Signal processing.

8.1. SIGNAL PROCESSING

We are continuously surrounded by signals, light signals, audio signals, electrical signals. Our experience of the world depends on them as they transmit the information about our everyday environment. A signal is understood as an alteration propagated in time and/or space, when a signal propagates only in time it will be expressed as $f(t)$; when it propagates only in space it will be expressed as $f(x)$; when it propagates in both, it will be expressed as $f(x,t)$. It is important to note that much of the mathematical formalism used in signal processing had been developed already during the creation of Acoustics as a field of study, particularly most formulations for continuous-time systems. As a general rule, signals are phenomena not well behaved in time, however, in the frequency domain, they can be easily handled. There are three common transforms, Fourier transform, Laplace transform and convolution. Two of them have already been explained in the section of Acoustics, now we will see Laplace transform. Laplace transform is commonly expressed as $\mathcal{L}$, it is a generalized form of the Fourier transform that takes a time function to the s plane (s-plane) $\mathcal{L}\{f\}: t \to s$. Although it is essentially a transform type $\mathcal{L} = T: \mathbb{R} \to \mathbb{R}$ Laplace transform (8.1) is a linear transformation, *i.e.*, $\mathcal{L}\{f + g\} = \mathcal{L}\{f\} + \mathcal{L}\{g\}$, defined as:

$$\mathcal{L}\{f(t)\}(s) = \int_0^\infty e^{-st} f(t) \; dt \tag{8.1}$$

[*]**Corresponding author Carlos Polanco:** Faculty of Sciences, Universidad Nacional Autónoma de México, México City, México; Tel: +01 55 5622 4858; Fax: +01 5556 4859; E-mail: polanco@unam.mx

Note 8.1. Laplace transform is an integral transform, where e^{-st} is the kernel.

Example 8.1. Transform function $f(t) = e^{-bt}t$ from its time domain $t \in \mathbb{R}$ to its domain in the s plane, where $s \in \mathbb{R}$.

Solution 8.1. Applying the transform as defined, $F(s) = \int_0^\infty e^{-st}e^{-bt}t \, dt$, manipulating it we have $F(s) = \int_0^\infty e^{-st-bt}t \, dt$ factoring t, $F(s) = \int_0^\infty e^{-(s+b)t}t \, dt$ integrating by parts $u = t$ then $du = dt$ and $dv = e^{-(s+b)t}dt$ and $v = -\frac{e^{-(s+b)t}}{(s+b)}$, hence $F(s) = \int_0^\infty e^{-(s+b)t}t \, dt = -\frac{e^{-(s+b)t}}{(s+b)}t - \int_0^\infty -\frac{e^{-(s+b)t}}{(s+b)} \, dt$ that is $F(s) = -\frac{e^{-(s+b)t}}{(s+b)}t + \frac{1}{(s+b)}\int_0^\infty e^{-(s+b)t} \, dt$ factoring $F(s) = \frac{1}{(s+b)}\left[-e^{-(s+b)t}t + \int_0^\infty e^{-(s+b)t}dt\right]$, integrating $F(s) = \frac{1}{(s+b)}\left[-e^{-(s+b)t}t - \frac{e^{-(s+b)t}}{(s+b)}\right]$, therefore, $F(s) = -\frac{1}{(s+b)}\left[e^{-(s+b)t}t + \frac{e^{-(s+b)t}}{(s+b)}\right]$ evaluating with the boundary condition of the integration we have $F(s) = -\frac{1}{(s+b)}[(1-0) + \frac{1}{(s+b)}(1-0)]$. So $F(s) = -\frac{1}{(s+b)}[1 + \frac{1}{(s+b)}]$.

CHAPTER 9

Epidemiology

Carlos Polanco[*]

Faculty of Sciences, Universidad Nacional Autónoma de México, México

Abstract: Epidemiological surveillance is a crucial field of Medicine, an early warning of an outbreak depends on the efficiency of a surveillance system. At present, the methodology to give an early warning has not changed, it is based on a minimum number of verified cases tested in a laboratory. However, there are already valuable contributions to assume that in the near future, this methodology will have a stochastic component.

Keywords: Deterministic models, Electronic devices, Epidemiology, Hidden Markov models, Real time monitoring, Translational medicine.

9.1. BACKGROUND

Epidemiological surveillance methods are designed to contain and minimize the number of infected cases and mortality. However, the increase and improvement of communication routes between towns, villages, and cities throughout the world, particularly in Africa, begin to compromise the effectiveness of these systems.

9.2. CUSUM MODEL

This model considers a minimum number of cases of the disease confirmed by microbiological testing. When this minimum is reached, depending on the disease, the outbreak warning is issued. The metrics model is a linear transformation. The advantage of this model is that it is based on confirmed cases, which means the population is facing a new outbreak. However, the warning may arrive when the disease has already affected the population. This algorithm is linear because $T(x_i + y_i) = T(x_i) + T(y_i)$, and $T(\alpha x_i) = \alpha T(x_i), i \in \mathbb{N}$.

Example 9.1. The 2014 Ebola outbreak has been the largest in West Africa. It was the first Ebola epidemic the world has ever known [CDC, 2017c] § Current Outbreaks. The first confirmed cases were in May 2014 but it was until November

[*]**Corresponding author Carlos Polanco:** Faculty of Sciences, Universidad Nacional Autónoma de México, México City, México; Tel: +01 55 5622 4858; Fax: +01 5556 4859; E-mail: polanco@unam.mx

2014 that the outbreak was confirmed. Total Cases (Suspected, Probable, and Confirmed) 28616. Laboratory-Confirmed Cases 15227. Total Deaths 11310 as of April 13, 2016. (Updated April 13, 2016) [CDC, 2017b]. On March 29, 2016, WHO terminated the Public Health Emergency of International Concern for the Ebola outbreak in West Africa [CDC, 2017a].

9.3. RANDOM MODEL

The Markov model [Eddy, 1996b] considers variables not directly associated with the epidemic event, such as: demographic, geographic, and climatic variables, access to piped water, among others. These model makes possible the prediction, in time and place, of the emergence of an outbreak. This algorithm, as all Markovian models, is nonlinear, *i.e.* $T(x_i + y_i) \neq T(x_i) + T(y_i)$, and $T(\alpha x_i) \neq \alpha T(x_i), i \in \mathbb{N}$.

Example 9.2. The models named Overcrowd Severe Respiratory Disease Index (OSRDI) [Polanco *et al.*, 2013e] and Índice de Saturación Modificado (ISM) [Polanco *et al.*, 2013d], are based on a Markov model. They can predict an outbreak based on the demand for hemodynamic monitors and mechanical ventilators used out of the seasonal period.

9.4. REAL TIME MONITORING

In Epidemiology, surveillance in real-time is crucial due to the accelerated interconnection between towns and cities that encourages the spread of a disease [Polanco, 2015]. In the near future, electronic sensors, bio-nano-robots implanted in individuals, and improvements in satellite communications will enable the collection of large amounts of biometric data that will be transmitted directly to epidemiological surveillance systems. These systems will need to have a predictive component that weights the number of cases giving the probability of having an outbreak. A model capable of doing so will have to be a nonlinear model.

Example 9.3. The natural reservoir located in Africa between the tropics of Cancer and Capricorn have the ideal conditions for the highest diversity of microorganisms. The spread of a disease in such conditions would be represented $T(x_i) = \alpha T(x_i) \neq T(x_i)$ thus, the spread would be a nonlinear factor.

CHAPTER 10

Genetics

Carlos Polanco* and Carlos Ignacio Herrera

Faculty of Sciences, Universidad Nacional Autónoma de México, México

Abstract: Genetics focuses on the study of the hereditary transmission of the anatomical, cytologic, and functional characters from parents to offspring. This chapter reviews the linear transformations that can be related to that and their extensions.

Keywords: Bayes's theorem, Genetics.

10.1. INTRODUCTION

Genetics is the branch studying genes and the hereditary characters of living organisms. From the mathematical point of view, a gene can be represented as a letter and a combination of them, as an ordered pair, can result in a range of possibilities. For this example, we will use some concepts of probability and their application to this branch of science.

Definition 10.1. A triplet $(\Omega, A, \mathbb{P})$ is a probability space, if to a set of events Ω and A is associated a σ -Algebra. There is a function $\mathbb{P} : \Omega \to [0,1]$ such that: (i) $\mathbb{P}(\emptyset) = 0$. (ii) Given $A \in \Omega \Rightarrow A^c \in \Omega$ and $\mathbb{P}(A^c) = 1 - \mathbb{P}(A)$. (iii) Given $A, B \in \Omega$ events, then $\mathbb{P}(A \cup B) = \mathbb{P}(A) + \mathbb{P}(B)$ if $(A \cup B) = \emptyset$.

Definition 10.2. Given a set of events Ω, if $A \in \Omega$ then

$$\mathbb{P}(A) = \frac{Number of favorable cases from A}{total cases}. \tag{10.1}$$

Note 10.1. This is known as the "Classical probability definition".

Property 10.1. Given $A, B \in \Omega$ events, then

$$\mathbb{P}(A \cap B) = \mathbb{P}(A)\mathbb{P}(B) \text{if}(A \cap B) = \emptyset. \tag{10.2}$$

---

*Corresponding author Carlos Polanco:** Faculty of Sciences, Universidad Nacional Autónoma de México, México City, México; Tel: +01 55 5622 4858; Fax: +01 5556 4859; E-mail: polanco@unam.mx

Note 10.2. This property is known as "Independence of events". Intuitively, it is said that two events are independent if the result of any of them does not affect the other.

Property 10.2. It can be said that given $A, B \in \Omega$ events, if

$$\mathbb{P}(B) > 0 \text{ then } \mathbb{P}(A|B) = \frac{\mathbb{P}(A \cap B)}{\mathbb{P}(B)} \tag{10.3}$$

$\mathbb{P}(A|B)$ is read as "The probability of A given B", this is known as the conditional probability of an event over another.

The intuitive perception of this property of events is to measure how feasible is an event to happen conditioned to another that has already happened. So far, we have defined a probability function and some of its properties under certain conditions; however, one of the most important properties in this theoretical framework is the **Bayes' theorem**.

10.2. BAYES THEOREM

Set $(A) \in \Omega$ can be expressed as follows: $(A) = (A \cap B) \cup (A \cap B^c)$ where $\mathbb{P}(A) = \mathbb{P}(A \cap B) + \mathbb{P}(A \cap B^c)$. That is

$$\mathbb{P}(A) = \mathbb{P}(A|B)\mathbb{P}(B) + \mathbb{P}(A|B^c)\mathbb{P}(B^c)$$

for any set $B \in \Omega$. The importance of this equation is to determine the probability of an event A conditioned to another B, apparently unrelated. This will be very useful for an event whose probability is not evident, as it will make it easy to detect it.

Example 10.1. Genes associated with albinism are denoted by (A) and (a). Only individuals receiving gene (a) from both parents have this condition. Individuals with both genes (A, a) are called carriers because they do not have the trait but they can inherit it to their descendants. Suppose a normal couple has two children, one of them is albino, suppose also that the non-albino child marries a carrier. (i) What is the probability their first descendant is albino? (ii) What is the probability their second descendant is albino since the first one was not? [Sheldon, 1976].

Solution 10.1. (i) To solve this part of the problem we must be aware the parent may or may not be a carrier. To visualize this, we will define the following: Be $A=$ *"The event where the parent is a carrier"* the fact of not being albino as his brother, will make him have the following combination of genes: $\Omega = \{(a,A),(A,a),(A,A)\}$. Considering that the only pair of genes expressing albinism is pair (a,a), and since he does not have this condition, it is assumed he does not have this pair of genes. On the other hand, by the classical probability definition, we have: $\mathbb{P}(A) = \frac{2}{3}, \mathbb{P}(A^c) = \frac{1}{3}$. Furthermore, be $B =$ *"the event where the first descendant is albino"*, using Bayes theorem we have: $\mathbb{P}(B) = \mathbb{P}(B|$ *that the parent may or may not be a carrier*) $\mathbb{P}(B|A)\mathbb{P}(A) + \mathbb{P}(B|A^c)\mathbb{P}(A^c)$. If the parent is a carrier, his descendant will have any of the following combination of genes: $\chi = \{(a,a),(a,A),(A,a),(A,A)\}$where $\mathbb{P}(B) = \frac{1}{4}$. If the parent is not a carrier, then the descendant will have any of the following combination of genes: $\chi = \{(a,A),(A,a),(A,A)\}$ where $\mathbb{P}(B) = 0$. Thus $\mathbb{P}(B) = \mathbb{P}(B|A)\mathbb{P}(A) + \mathbb{P}(B|A^c)\mathbb{P}(A^c) = \frac{1}{4} \cdot \frac{2}{3} + 0 \cdot \frac{1}{3} = \frac{2}{12} = \frac{1}{6}$. (ii) To solve the second part of the problem: Be $C =$ *"The event where the second descendant is albino"*, where $\mathbb{P}(C|B^c) = \frac{\mathbb{P}(C \cap B^c)}{\mathbb{P}(B^c)}$. At this point, it might not be clear how to calculate $\mathbb{P}(C|B^c)$. However, we can condition whether "the parent may or may not be a carrier, therefore, we would have that:

$$\frac{\mathbb{P}(C \cap B^c)}{\mathbb{P}(B^c)} =$$

$$\frac{\mathbb{P}(C \cap B^c|theparentmayormaynotbeacarrier)}{\mathbb{P}(B^c)} =$$

$$= \frac{\mathbb{P}(C|A)\mathbb{P}(B^c|A)\mathbb{P}(A)+\mathbb{P}(C|A^c)\mathbb{P}(B^c|A^c)\mathbb{P}(A^c)}{\mathbb{P}(B^c)}.$$

In the same way than the first question, if the parent is a carrier we have the following combination of genes for the second descendant: $\Psi = \{(a,a),(a,A),(A,a),(A,A)\}$. And if the parent is not a carrier:

$$\Psi = \{(a,A),(A,a),(A,A)\}.$$

Then:

$$\mathbb{P}(C|B^c) = \frac{\frac{1}{4} \cdot \frac{3}{4} \cdot \frac{2}{3}+0 \cdot \frac{3}{3} \cdot \frac{1}{3}}{\frac{5}{6}} = \frac{\frac{1}{4} \cdot \frac{3}{4} \cdot \frac{2}{3}}{\frac{5}{6}} = \frac{\frac{1}{8}}{\frac{5}{6}} = \frac{3}{20}.$$

This example has shown how we can predict events based on simple assumptions and probabilistic properties. In this case, the Bayes theorem was key and the main tool used. It is important to note the nonlinear character of the Bayes theorem since its outlook is more of a conjunction and although it is possible to weight and adapt a set space to meet the properties of a vector space, it is not possible to ensure that the linear properties in the definitions mentioned above will be met. However, in the field of probability, it is feasible to define operators as linear operators.

10.3. GENOTYPES

In the previous example 2, it was shown how the probability of events related to albinism can be estimated, this was calculated based on the properties of probability and information about how genes are transmitted. The next example ? Ch. 11, pp. $733 - 735$, will show another method, this time based on linear algebra. Many problems in nature are based on the system approach $A\bar{x} = \lambda\bar{x}$ where A is a matrix from $(n \times n)$, $\bar{x} \in \mathbb{R}^n$ and $\lambda \in \mathbb{R}$. This system is solved for vectors $\bar{x}$ such that $(A - \lambda I)\bar{x} = \bar{0}$.

Definition 10.3. Be λ that satisfies the equation $(A - \lambda I)\bar{x} = \bar{0}$ then λ is an egenvector of the system. In the same way, for each λ there is $\bar{v} \neq 0$ that satisfies the equation $\bar{v}$ called associated-eigenvector of λ. To determine the eigenvalues that satisfy this equality, we take $det(A - \lambda I) = 0$, the equation is known as *characteristic equation*. Finding the eigenvectors of matrix A is similar to the diagonalization of a matrix.

Definition 10.4. If A is a matrix of $(n \times n)$, A is diagonizable if there is a Q matrix of (nxn) invertible such that $A = QDQ^{-1}$ where $Q = (v_1 \quad v_2 \quad v_3 \quad \cdots \quad v_n)$, v_i is an eigenvector $\forall i \in \{1 \cdots n\}$ and

$$D = \begin{pmatrix} \lambda_1 & 0 & 0 & \cdots & 0 \\ 0 & \lambda_2 & 0 & \cdots & 0 \\ 0 & 0 & \lambda_3 & \cdots & 0 \\ \vdots & \vdots & \vdots & \ddots & 0 \\ 0 & 0 & 0 & 0 & \lambda_n \end{pmatrix} \text{ with } \lambda_i \text{ eigenvalues } \forall i \in \{1 \cdots n\}. \textbf{(10.4)}$$

Example 10.2. Suppose a farmer has a large population of plants with a distribution of genotypes $(a, a), (a, A), (A, A)$. The farmer wants to start a plant-breeding program in which each plant is always fertilized with a plant of the genotype (a, A) that can later be replaced by one of its offsprings. Express the distribution of the three possible genotypes after any number of generations.

Solution 10.2. Consider the following table showing the probabilities to obtain any of the three genotypes related to the parents:

Be

$$a_n = \text{the plants type (A,A) in the same generation}$$

$$b_n = \text{the plants type (A,a) in the same generation}$$

$$c_n = \text{the plants type (a,a) in the same generation}$$

Parents Offspring	(A,A) (A,A)	(A,A) (A,a)	(A,A) (a,a)	(A,a) (A,a)	(A,a) (a,a)	(a,a) (a,a)
(A,A)	1	$\frac{1}{2}$	0	$\frac{1}{4}$	0	0
(A,a)	0	$\frac{1}{2}$	1	$\frac{1}{2}$	$\frac{1}{2}$	0
(a,a)	0	0	0	$\frac{1}{4}$	$\frac{1}{2}$	1

where:

$$a_n + b_n + c_n = 1, \forall n \in \mathbb{N}$$

Thus, the equations can be deduced in the following way:

$$a_n = \frac{1}{2} a_{n-1} + \frac{1}{4} b_{n-1}$$

$$b_n = \frac{1}{2} a_{n-1} + \frac{1}{2} b_{n-1} + \frac{1}{2} c_{n-1}$$

$$c_n = \frac{1}{4} b_{n-1} + \frac{1}{2} c_{n-1}.$$

This equations can be expressed as matrix $A\bar{x}^{n-1} = \bar{x}^n$ where

$$A = \begin{pmatrix} \dfrac{1}{2} & \dfrac{1}{4} & 0 \\[2mm] \dfrac{1}{2} & \dfrac{1}{2} & \dfrac{1}{2} \\[2mm] 0 & \dfrac{1}{4} & \dfrac{1}{2} \end{pmatrix},$$

$$\bar{x}^{n-1} = \begin{pmatrix} a_{n-1} \\ b_{n-1} \\ c_{n-1} \end{pmatrix},$$

and

$$\bar{x}^{n} = \begin{pmatrix} a_{n} \\ b_{n} \\ c_{n} \end{pmatrix}$$

If $A\bar{x}^{n-1} = \bar{x}^{n}$, then $A^2\bar{x}^{n-2} = \bar{x}^{n} = A^3\bar{x}^{n-3} = \cdots = A^n\bar{x}^0$. Thus, we have an expression based on the initial plant population $\bar{x}^0$ and matrix A raised to the n power. Note that as stated in the definitions, $A = QDQ^{-1}$ where $A^n = QD^nQ^{-1}$. To calculate the parts of this equation we have (1) eigenvalues, and (2) eigenvectors respectively.

(1) $det(A - \lambda I) = 0$

$$det \begin{pmatrix} \dfrac{1}{2} - \lambda & \dfrac{1}{4} & 0 \\[2mm] \dfrac{1}{2} & \dfrac{1}{2} - \lambda & \dfrac{1}{2} \\[2mm] 0 & \dfrac{1}{4} & \dfrac{1}{2} - \lambda \end{pmatrix} = 0$$

$$(\tfrac{1}{2} - \lambda)((\tfrac{1}{2} - \lambda)^2 - \tfrac{1}{4}) = 0$$

$$\lambda_1 = \frac{1}{2}; \lambda_2 = 1; \lambda_3 = 0$$

(2) For $\lambda = \frac{1}{2}$

$$(A - \lambda I)\bar{x} = \bar{0}$$

$$\begin{pmatrix} \frac{1}{2} - \frac{1}{2} & \frac{1}{4} & 0 \\ \frac{1}{2} & \frac{1}{2} - \frac{1}{2} & \frac{1}{2} \\ 0 & \frac{1}{4} & \frac{1}{2} - \frac{1}{2} \end{pmatrix} \begin{pmatrix} x \\ y \\ z \end{pmatrix} = \begin{pmatrix} 0 \\ 0 \\ 0 \end{pmatrix}$$

$$\begin{pmatrix} 0 & \frac{1}{4} & 0 \\ \frac{1}{2} & 0 & \frac{1}{2} \\ 0 & \frac{1}{4} & 0 \end{pmatrix} \begin{pmatrix} x \\ y \\ z \end{pmatrix} = \begin{pmatrix} 0 \\ 0 \\ 0 \end{pmatrix}.$$

Solving the system of equations, we have that the vector satisfying the equation is $\bar{v}_1 = \begin{pmatrix} 1 \\ 0 \\ -1 \end{pmatrix}$, using the same procedure for $\lambda = 1$, and $\lambda = 0$ we have $\bar{v}_2 = \begin{pmatrix} 1 \\ 2 \\ 1 \end{pmatrix}$, and $\bar{v}_3 = \begin{pmatrix} 1 \\ -2 \\ 1 \end{pmatrix}$. So $Q = (v_1 \quad v_2 \quad v_3) = \begin{pmatrix} 1 & 1 & 1 \\ 0 & 2 & -2 \\ -1 & 1 & 1 \end{pmatrix}$, obtaining its inverse

we have $Q^{-1} = \begin{pmatrix} \frac{1}{2} & 0 & -\frac{1}{2} \\ \frac{1}{4} & \frac{1}{4} & \frac{1}{4} \\ \frac{1}{4} & -\frac{1}{4} & \frac{1}{4} \end{pmatrix}$.

Further $D = \begin{pmatrix} \lambda_1 & 0 & 0 \\ 0 & \lambda_2 & 0 \\ 0 & 0 & \lambda_3 \end{pmatrix} \Rightarrow D^n = \begin{pmatrix} \lambda_1^n & 0 & 0 \\ 0 & \lambda_2^n & 0 \\ 0 & 0 & \lambda_3^n \end{pmatrix},$

hence:

$$A^n = QD^nQ^{-1} \Rightarrow A^n = \begin{pmatrix} 1 & 1 & 1 \\ 0 & 2 & -2 \\ -1 & 1 & 1 \end{pmatrix} \begin{pmatrix} \lambda_1^n & 0 & 0 \\ 0 & \lambda_2^n & 0 \\ 0 & 0 & \lambda_3^n \end{pmatrix} \begin{pmatrix} \dfrac{1}{2} & 0 & -\dfrac{1}{2} \\ \dfrac{1}{4} & \dfrac{1}{4} & \dfrac{1}{4} \\ \dfrac{1}{4} & -\dfrac{1}{4} & \dfrac{1}{4} \end{pmatrix}.$$

making the product:

$$A^n = \begin{pmatrix} 1 & 1 & 1 \\ 0 & 2 & -2 \\ -1 & 1 & 1 \end{pmatrix} \begin{pmatrix} (\frac{1}{2})\lambda_1^n & 0 & -(\frac{1}{2})\lambda_1^n \\ (\frac{1}{4})\lambda_2^n & (\frac{1}{4})\lambda_2^n & (\frac{1}{4})\lambda_2^n \\ (\frac{1}{4})\lambda_3^n & -(\frac{1}{4})\lambda_3^n & (\frac{1}{4})\lambda_3^n \end{pmatrix}$$

$$A^n = \begin{pmatrix} (\frac{1}{2})\lambda_1^n + (\frac{1}{4})\lambda_2^n + (\frac{1}{4})\lambda_3^n & (\frac{1}{4})\lambda_2^n - (\frac{1}{4})\lambda_3^n & -(\frac{1}{2})\lambda_1^n + (\frac{1}{4})\lambda_2^n + (\frac{1}{4})\lambda_3^n \\ (\frac{1}{2})\lambda_2^n - (\frac{1}{2})\lambda_3^n & (\frac{1}{2})\lambda_2^n + (\frac{1}{2})\lambda_3^n & (\frac{1}{2})\lambda_2^n - (\frac{1}{2})\lambda_3^n \\ (\frac{1}{2})\lambda_1^n + (\frac{1}{4})\lambda_2^n + (\frac{1}{4})\lambda_3^n & (\frac{1}{4})\lambda_2^n - (\frac{1}{4})\lambda_3^n & (\frac{1}{2})\lambda_1^n + (\frac{1}{4})\lambda_2^n + (\frac{1}{4})\lambda_3^n \end{pmatrix}$$

$$A^n = \begin{pmatrix} (\frac{1}{2})(\frac{1}{2})^n + (\frac{1}{4})(1)^n + (\frac{1}{4})(0)^n & (\frac{1}{4})(1)^n - (\frac{1}{4})(0)^n & -(\frac{1}{2})(\frac{1}{2})^n + (\frac{1}{4})(1)^n + (\frac{1}{4})(0)^n \\ (\frac{1}{2})(1)^n - (\frac{1}{2})(0)^n & (\frac{1}{2})(1)^n + (\frac{1}{2})(0)^n & (\frac{1}{2})(1)^n - (\frac{1}{2})(0)^n \\ (\frac{1}{2})(\frac{1}{2})^n + (\frac{1}{4})(1)^n + (\frac{1}{4})(0)^n & (\frac{1}{4})(1)^n - (\frac{1}{4})(0)^n & (\frac{1}{2})(\frac{1}{2})^n + (\frac{1}{4})(1)^n + (\frac{1}{4})(0)^n \end{pmatrix}$$

$$A^n = \begin{pmatrix} (\frac{1}{2})^{n+1} + (\frac{1}{4}) & (\frac{1}{4}) & -(\frac{1}{2})^{n+1} + (\frac{1}{4}) \\ (\frac{1}{2}) & (\frac{1}{2}) & (\frac{1}{2}) \\ (\frac{1}{2})^{n+1} + (\frac{1}{4}) & (\frac{1}{4}) & (\frac{1}{2})^{n+1} + (\frac{1}{4}) \end{pmatrix}.$$

We have already calculated all elements required to find the equations that satisfy system $A^n \bar{x}^0 = \bar{x}^n$. Then:

$$A^n \bar{x}^0 = \bar{x}^n \Rightarrow \begin{pmatrix} (\frac{1}{2})^{n+1} + (\frac{1}{4}) & (\frac{1}{4}) & -(\frac{1}{2})^{n+1} + (\frac{1}{4}) \\ (\frac{1}{2}) & (\frac{1}{2}) & (\frac{1}{2}) \\ (\frac{1}{2})^{n+1} + (\frac{1}{4}) & (\frac{1}{4}) & (\frac{1}{2})^{n+1} + (\frac{1}{4}) \end{pmatrix} \begin{pmatrix} a_0 \\ b_0 \\ c_0 \end{pmatrix} = \begin{pmatrix} a_n \\ b_n \\ c_n \end{pmatrix}$$

$\Rightarrow$

$$a_0((\tfrac{1}{2})^{n+1} + (\tfrac{1}{4})) + b_0(\tfrac{1}{4}) + c_0(-(\tfrac{1}{2})^{n+1} + (\tfrac{1}{4})) = a_n$$

$$a_0(\tfrac{1}{2}) + b_0(\tfrac{1}{2}) + c_0(\tfrac{1}{2}) = b_n$$

$$a_0((\tfrac{1}{2})^{n+1} + (\tfrac{1}{4})) + b_0(\tfrac{1}{4}) + c_0((\tfrac{1}{2})^{n+1} + (\tfrac{1}{4})) = c_n; \forall n \geq 1$$

These are the equations required to solve this problem. Finally, if we take $n \to \infty$, it will be observed the distribution tends to the following:

$$a_0(\tfrac{1}{4}) + b_0(\tfrac{1}{4}) + c_0(\tfrac{1}{4}) = a_n$$

$$a_0(\tfrac{1}{2}) + b_0(\tfrac{1}{2}) + c_0(\tfrac{1}{2}) = b_n$$

$$a_0(\tfrac{1}{4}) + b_0(\tfrac{1}{4}) + c_0(\tfrac{1}{4}) = c_n$$

In this practical example can be seen the application developed directly from linear algebra calculating eigenvalues and eigenvectors, *i.e.* the solution of the problem for the system $(A - \lambda I)\bar{x} = \bar{0}$ was entirely a linear transformation.

CHAPTER 11

Optics

Carlos Polanco[*]

Faculty of Sciences, Universidad Nacional Autónoma de México, México

Abstract: This scientific discipline focuses on the study of the behavior and properties of light such as the energy of a waveform or an electromagnetic wave. The concept of light is geometrically represented as a set of rays travelling in straight line, bending when it passes through or reflects from a surface. The study of the reflection and refraction of rays in different forms is called Optics [Cornejo-Rodríguez and Urcid-Serrano, 2005] p. 8.

Keywords: Malus law, Spherical aberration.

11.1. ABERRATION THEORY

There is a number of simplifications for the study of the main properties and characteristics of optical systems [Jenkins and White, 1937] pp. 152-157, among them, the assumption that the sine of an angle is equal to the value of its radians. Thus, we have first order Gaussian optics. When the sine of an angle is substituted by the two first terms of a series, we have a third-order theory. More terms of a series are included for higher-order theories.

11.2. SNELL'S LAW

In the study of the propagation of light, it is important to know the phenomenon that occurs when a beam of light passes from one medium to another with different refractive index. Snell's law is a mathematical expression that determines how much a beam of light diverts when passing from one medium to another as described below:

$$n\sin I = n'\sin I' \tag{11.1}$$

Where n and n' are the indices of refraction of the incident medium and the refractive or transmission medium, respectively; and I and I' are the incidence and

---

[*]**Corresponding author Carlos Polanco:** Faculty of Sciences, Universidad Nacional Autónoma de México, México City, México; Tel: +01 55 5622 4858; Fax: +01 5556 4859; E-mail: polanco@unam.mx

refraction angles. The angle is measured from the normal to the plane of incidence $I \in [0, \pi/2]$. Snell's law is a function of the angle of incidence I and we know that:

$$\sin(I + J) \neq \sin I + \sin J, \tag{11.2}$$

therefore, this is not a linear transformation.

11.3. SPHERICAL ABERRATION

Example 11.1. From the Gauss equation, we know that:

$$n'L' - nL = n' - nr + n\sin Ir(\sin I - \sin U)(1 - \sin I - \sin U \sin I' - \sin U')$$

We also know,

$$\sin I \sin I - \sin U = L - rL$$

Thus, by substitution we have,

$$n'L' - nL = n' - nr + n(L - r)rL(1 - \sin I - \sin U \sin I' - \sin U')$$

Using the trigonometric relation,

$$\sin I - \sin U = 2\sin(I - U2)\cos(I + U2)$$

We have

$$n'L' - nl = n' - nr + n(L - r)rL[\cos(I' + U'2) - \cos(I + U2)\cos(I' + U'2)]$$

11.4. MALUS LAW

When we speak about the polarization of light [Jenkins and White, 1937] pp. 503,504, we refer to the direction of the electric vector in relation to the polarization plane. If the electric vector, which is a time and space function, only varies in a sinusoidal form we say the light is linearly polarized. In natural light, direction and magnitude are random functions of time and space. If we make a beam of polarized light pass through an analyzer, which is a filter with an axis parallel to its plane, in such a way that the incident beam can pass without attenuation only if the axis of the polarizer is in the plane of oscillation of the incident beam, then by turning the analyzer, only the projection of the electric vector on the axes of the analyzer will go through. We express the amplitude as:

$$E_T = E\cos\theta \tag{11.3}$$

Thus, the irradiance will be:

$$I_T = I\cos^2\theta \tag{11.4}$$

Where I_T is the transmitted irradiance, I is the incident irradiance and θ is the angle of deflection of the analyzer. The irradiance transmitted is the function of $\theta \in [0, \pi]$ and we know that $\sin(\theta + \varphi) \neq \sin\theta + \sin\varphi$, therefore, this is not a linear transformation.

CHAPTER 12

Probability

Carlos Polanco[*] and Alma Fernanda Sánchez

Faculty of Sciences, Universidad Nacional Autónoma de México, México

Abstract: Experimental verification of different scientific hypotheses requires one or more statistical tests. These tests assume that the samples studied behave according to different distributions: Normal, Binomial and Exponential. In this chapter, we will study the linear or nonlinear nature of these distributions.

Keywords: Binomial distribution, Exponential distribution, Normal distribution.

12.1. BACKGROUND

12.2. BINOMIAL DISTRIBUTION

Binomial distribution [Casella and Berger, 2002; Rincón, 2014] is a discrete distribution that results from a sequence of independent events, each one with a Bernoulli distribution. For instance, it is said that $X \sim Bin(n,p)$ (12.1) if X has a number of successes in n as independent trials and it has a density function:

$$f(x) = P(X = x) = \begin{cases} \dbinom{n}{x} p^x (1-p)^{n-x} & : x \in \mathbb{Z}^+ \\ 0 & : \text{otherwise} \end{cases} \qquad (12.1)$$

where n represents the number of tests carried out, p the probability of success or that the event occurs, and $1 - p$ the probability of "failure" or that the event does not occur).

Example 12.1. Suppose we apply a test with ten questions whose answers can be YES or NO. (i) Find the probability to obtain 5 hits. (ii) Find the probability to obtain some success.

---

[*]**Corresponding author Carlos Polanco:** Faculty of Sciences, Universidad Nacional Autónoma de México, México City, México; Tel: +01 55 5622 4858; Fax: +01 5556 4859; E-mail: polanco@unam.mx

Solution 12.1. (i) $n = 10$ (10 questions to answer). $p = 12$ y $q = 1 - p = \frac{1}{2}$, because there are two possible answers to each question Yes or NO. As it is required to obtain 5 hits then $X = 5 \Rightarrow P(X = 5) = n_x\, p^x(1 - p)^{n-x} \Rightarrow$ $10_5\, (\frac{1}{2})^5(\frac{1}{2})^{10-5} = 252(\frac{1}{32})(\frac{1}{32}) = \frac{252}{1024} \approx 0.2461$. (ii) We want to know the probability to obtain some success, this is equivalent to $X \geq 1 \Rightarrow P(X \geq 1) = 1 - P(X < 1), 1 - P(X = 0) = 1 - 10_0\, (\frac{1}{2})^0(\frac{1}{2})^{10-0} = 1 - (1)(1)(\frac{1}{1024}) = (\frac{1023}{1024}) \approx 0.9990$.

If we have two independent random variables $X \sim Bin(n, p)$ and $Y \sim Bin(m, p)$. How is their sum distributed?. This is solved by a convolution. Be two given functions f and g, we know that $f_Z(z) = \sum_{x=0}^{z} f_X(x) f_Y(z - x)$, if $y = z - x \Rightarrow$ $f_Z(z) = \sum_{x=0}^{z} n_x\, p^x(1 - p)^{n-x} m_{z-x}\, p^{z-x}(1 - p)^{m-(z-x)} =$ $\sum_{x=0}^{z} n_x\, m_{z-x}\, p^x p^{z-x}(1 - p)^{n-x}(1 - p)^{m-z+x} = \sum_{x=0}^{z} n_x\, m_{z-x}\, p^z(1 - p)^{n+m-z}$. Using Vandermonde's identity get function Z with binomial distribution parameters $(m + m, p)$. So $Z \sim Bin(m + n, p)$ where $Z = X + Y$. Thus, the sum of independent binomial random variables is a nonlinear transformation because $T(X, Y) \neq T(X) + T(Y)$. Therefore, the convolution operator is nonlinear.

Note 12.1. Vandermonde's identity is $\sum_{x=0}^{z} n_x\, m_{z-x} = n + m_z$ where $m, n, r \in \mathbb{N}$.

12.3. EXPONENTIAL DISTRIBUTION

Exponential distribution is a continuous distribution to measure the time between events. It can model "failure-rate" problems and serves as a model for the study of inter-arrival time problems. It can also measure the time gap before a *poisson event occurs and it models the time span between two consecutive poisson events occurring independently at a constant frequency.*

Example 12.2. Suppose the lifespan of a circuit obeys the exponential law $\lambda = \frac{1}{1000}$. A company producing these circuits wants to give a product guarantee. How many hours can the company guarantee the product to cover its lifespan at 95%?

Solution 12.2. With the available information, we know that $X \sim exp(\lambda = \frac{1}{1000})$

$$f(x) = \frac{1}{1000}e^{-\frac{x}{1000}} \quad ; \quad x > 0$$

So we calculate $P(x \geq) = 0.95$ and solve for k

$$P(x \geq k) = 1 - P(x < k)$$

$$P(x < k) = \int_0^k \frac{1}{1000}e^{-\frac{x}{1000}}dx$$

$$u = -\frac{x}{1000} \quad du = -\frac{dx}{1000}$$

$$= -\frac{1000}{1000}\int_0^k e^u du$$

$$= -\int_0^k e^u du$$

$$= -(e^u|_0^k)$$

$$= (-e^{-\frac{x}{1000}}|_0^k)$$

$$= 1 - e^{-\frac{k}{1000}}.$$

$$P(x \geq k) = 1 - (1 - e^{-\frac{k}{1000}})$$

$$= e^{-\frac{k}{1000}}$$

$$= 0.95$$

$$-\frac{k}{1000} = \ln 0.95$$

$$k = -1000\ln 0.95$$

$$k = 51.29 \text{ hours}$$

Considering it is an important distribution used, let us study its linearity. Suppose we have a random variable X exponentially distributed with parameter λ, *i.e.*, $(x \sim exp(\lambda))$ and we want to calculate its product with any positive scalar, without loss of generality so let us take λ as scalar. On one hand, if we want to calculate λX

$$f(\lambda x) = \lambda e^{-\lambda(\lambda x)}$$

$$= \lambda e^{-\lambda^2 x} \tag{12.1}$$

On the other, in order to be linear (1) will have to be equal to $\lambda f(X)$, thus,

$$\lambda f(\lambda x) = \lambda(\lambda e^{-\lambda x})$$

$$= \lambda^2 e^{-\lambda x}$$

Therefore, it can be concluded this is a non-linear transformation since $f(\lambda x) \neq \lambda f(x)$

12.4. NORMAL DISTRIBUTION

Also known as a Gaussian distribution, normal distribution is widely used for the study of statistical phenomena as its peculiar shape and symmetry make it suitable to model these events, in a more accurate way and facilitate their study. It is also possible to approximate many of the known distributions to a normal distribution, under certain conditions. It can be said that X has a normal distribution if it has a density function given by:

$$f(x) = \frac{1}{\sqrt{2\pi\sigma^2}} e^{-\frac{(x-\mu)^2}{2\sigma^2}} \tag{12.2}$$

With $x \in R$, $\mu \in R$ y $\sigma^2 > 0$, where the last two are the parameters representing expectation and variance, respectively, and can be denoted as $X \sim N(\mu, \sigma^2)$. If the parameters are $\mu = 0$ and $\sigma^2 = 1$ it would be a standard normal distribution, which is a particular normal distribution.

Example 12.3. The lifetime of a bulb can be modelled by a random variable with normal distribution parameters $\mu = 20{,}000$hours and $\sigma = 500$hours. (i) What is the probability the bulb lasts more than 21000 hours? (ii) Is the normal distribution a linear transformation?

Solution 12.3. (i) We have a random variable X with normal distribution parameters $\mu = 2000$ y $\sigma = 500$.

$$\Rightarrow X \sim N(20000, (500)^2)$$

$$\Rightarrow X \sim N(20000, 25000)$$

We know that $X = 21000$ so we stardarized the function

$$\Rightarrow z = \frac{x - \mu}{\sigma} \sim N(0,1)$$

$$\Rightarrow \frac{21000 - 20000}{500} = 2 \sim N(0,1)$$

Then, to know the probability of $z > 2$.

$$\Rightarrow P(z > 2) = 1 - \phi(2)$$

$$= \phi(-2) = 0.02275$$

(ii) To study its linearity, we take variable X with normal distribution $(X \sim N(\mu, \sigma^2))$. In order to be a linear transformation, it has to comply with $f(cx) = cf(x)$ where $c \in R$ and $f(x + y) = f(x) + f(y)$. Using the first condition we can prove that the linearity is not met. On one hand we have:

$$f(cx) = \frac{1}{\sqrt{2\Pi\sigma^2}} e^{-\frac{(cx-\mu)^2}{2\sigma^2}}$$

On the other:

$$cf(x) = c\left(\frac{1}{\sqrt{2\Pi\sigma^2}} e^{-\frac{(x-\mu)^2}{2\sigma^2}}\right)$$

$$= \frac{c}{\sqrt{2\Pi\sigma^2}} e^{-\frac{(x-\mu)^2}{2\sigma^2}}$$

So, $f(cx) \neq cf(x)$. Therefore, it can be concluded this function is nonlinear.

CHAPTER 13

Vector Analysis

Carlos Polanco*

Faculty of Sciences, Universidad Nacional Autónoma de México, México

Abstract: The Stokes, Gauss, and Green theorems play an important role in Vector Analysis. The transformations involved simplify line and surface integrals. In this chapter, these transformations and integral theorems will be reviewed.

Keywords: Gauss theorem, Green theorem, Stokes theorem, Vector analysis.

13.1. BACKGROUND

13.2. STOKES THEOREM

Definition 13.1. Let S be an oriented surface [Marsden and Tromba, 2011] Ch. 8 in a space with boundary denoted by ∂S, and F a C^1 vector field on S. Then

$$\oint_{\partial S} F \cdot ds = \iint_S (\nabla \times F) \cdot dS. \tag{13.1}$$

Three transformations are required to simplify the integrals: $T: [u_0, u_1] \times [v_0, v_1] \subset \mathbb{R}^2 \to S \subset \mathbb{R}^3$; $C: [t_0, t_1] \subset \mathbb{R} \to \partial S \subset \mathbb{R}^3$. $\oint_{\partial S} F \cdot ds = \int_{t_0}^{t_1} F \circ C \cdot C' \, dt$; and $\iint_S (\nabla \times F) \cdot dS = \int_{v_0}^{v_1} \int_{u_0}^{u_1} (\nabla \times F) \circ T \cdot (T_u \times T_v) \, du \, dv$.

Example 13.1. Let ∂S curve be $\partial S = \{(x, y, z) \subset \mathbb{R}^3 \mid \sqrt{x^2 + y^2} \cap 1 - x - y\}$, and let vector field F be $F(x, y, z) = (-y^3, x^3, -z^3)$. Verify the Stokes theorem.

Solution 13.1. If $C: [0, 2\pi] \to \mathbb{R}^3, (\cos\theta, \sin\theta, 1 - \cos\theta - \sin\theta)$, and $T: [0,1] \times [0, 2\pi] \to \mathbb{R}^3, (r\cos\theta, r\sin\theta, 1 - r\cos\theta - r\sin\theta)$. (i) The line integral is $\oint_{\partial S} F \cdot ds = \int_{t_0}^{t_1} F \circ C \cdot C' \, dt = \int_0^{2\pi} (-\sin^3\theta, \cos^3\theta, -(1 - \cos\theta - \sin\theta)^3) \cdot (-\sin\theta, \cos\theta, \sin\theta - \cos\theta) \, d\theta = \int_0^{2\pi} \sin^4\theta + \cos^4\theta - (1 - \cos\theta - \sin\theta)^3 (\sin\theta - \cos\theta) \, d\theta = \int_0^{2\pi} \sin^4\theta d\theta + \int_0^{2\pi} \cos^4\theta \, d\theta - \int_0^{2\pi} (1 - \cos\theta - \sin\theta)^3 (\sin\theta - \cos\theta) \, d\theta = \frac{3}{2}\pi$. and (ii) the double integral $\iint_S (\nabla \times F) \cdot dS = \int_0^{2\pi} \int_0^1 (\nabla \times F) \circ T \cdot (T_r \times T_\theta) \, dr \, d\theta = \int_0^{2\pi} \int_0^1 3r^2 r \, dr \, d\theta = 6\pi[\frac{1}{4}r^4]_0^1 = \frac{3}{2}\pi$.

*Corresponding author **Carlos Polanco:** Faculty of Sciences, Universidad Nacional Autónoma de México, México City, México; Tel: +01 55 5622 4858; Fax: +01 5556 4859; E-mail: polanco@unam.mx

Note 13.1. Using $\int \cos^4\theta \, d\theta = \frac{\cos^{n-1}\theta\sin\theta}{n} + \frac{n-1}{n}[\frac{1}{2}(\theta + \sin\theta + \cos\theta)]$, and

$\int \sin^4\theta \, d\theta = \frac{\sin^{n-1}\theta\cos\theta}{n} + \frac{n-1}{n}[\frac{1}{2}(\theta - \sin\theta + \cos\theta)] \Rightarrow \int_0^{2\pi} \cos^4\theta \, d\theta =$

$[\frac{\cos^3\theta\sin\theta}{4}]_0^{2\pi} + \frac{3}{4}[\frac{1}{2}(\theta + \sin\theta + \cos\theta)]_0^{2\pi} = \frac{3}{8}(2\pi + 1) - \frac{3}{8} = \frac{3}{4}\pi.$

$\int_0^{2\pi} \sin^4\theta \, d\theta = [\frac{\sin^3\theta\cos\theta}{4}]_0^{2\pi} + \frac{3}{4}[\frac{1}{2}(\theta - \sin\theta + \cos\theta)]_0^{2\pi} = \frac{3}{8}(2\pi + 1) - \frac{3}{8} =$
$\frac{3}{4}\pi.$

Note 13.2. If $u = 1 - \cos\theta - \sin\theta \Rightarrow du = \sin\theta - \cos\theta$ $\int_0^{2\pi} (1 - \cos\theta - \sin\theta)^3(\sin\theta - \cos\theta) \, d\theta = \int_0^{2\pi} u^3 du = [\frac{1}{4}u^4]_0^0 = 0.$

Note 13.3. $\nabla \times F = (\frac{\partial}{\partial x}, \frac{\partial}{\partial y}, \frac{\partial}{\partial z}) \times (-y^3, x^3, -z^3) = (0,0,3x^2 + 3y^2) \Rightarrow (\nabla \times F) \circ$
$T = (0,0,3x^2 + 3y^2) \circ T(r,\theta) = (0,0,3r\cos^2\theta + 3r\sin^2\theta).$

Note   13.4. $T_r(r,\theta) = (\cos\theta, \sin\theta, -\cos\theta - \sin\theta)$,   and $T_\theta(r,\theta) = (-r\sin\theta, r\cos\theta, r\sin\theta - \cos\theta).$

13.3. GREEN THEOREM

Definition 13.2. Let S be an oriented surface [Marsden and Tromba, 2011] Ch. 8 in a plane with boundary denoted by ∂S, and F a C^1 vector field on S.

$$\oint_{\partial S} F \cdot ds = \iint_S (\nabla \times F) \cdot dS. \tag{13.2}$$

Three transformations are required to simplify the integrals: $T: [u_0, u_1] \times [v_0, v_1] \subset \mathbb{R}^2 \to S \subset \mathbb{R}^2$; $C: [t_0, t_1] \subset \mathbb{R} \to \partial S \subset \mathbb{R}^2$. $\oint_{\partial S} F \cdot ds = \int_{t_0}^{t_1} F \circ C \cdot C' \, dt$, and $\iint_S (\nabla \times F) \cdot dS = \int_{v_0}^{v_1} \int_{u_0}^{u_1} (\nabla \times F) \circ T \cdot (T_u \times T_v) |J(T)| \, du \, dv.$

Example 13.2. Let the ∂S curve be $\partial S = \{(x, y) \subset \mathbb{R}^2 \mid y = 2x \cap y = 2x - 2 \cap y = x \cap y = x + 1\}$, and let vector field F be $F(x, y) = (xy^3, y^2 + x)$. Verify the Green theorem.

Solution   13.2.   If $C_1: [0,2] \to \mathbb{R}^2, (t,t)$, $C_2: [0,1] \to \mathbb{R}^2, (2 + t, 2 + 2t)$, $C_3: [2,0] \to \mathbb{R}^2, (1 + t, 2 + t)$, $C_4: [1,0] \to \mathbb{R}^2, (t, 2t)$, and $T: [0,1] \times [-2,0] \subset \mathbb{R}^2 \to S \subset \mathbb{R}^2, (u - v, 2u - v)$. (i) The line integral is $\oint_{\partial S_i} F \cdot ds = \int_{t_0}^{t_1} F \circ C_i \cdot C_i, dt$ where $i = 1,4 = -\frac{2458}{15}$, (ii) the double integral is $\iint_S (\nabla \times F) \cdot dS = \int_{-2}^0$

$\int_0^1 (\nabla \times F) \circ T \;\cdot\; (T_u \times T_v) \, |J(T)| \; du \, dv = \int_{-2}^0 \int_0^1 13(u - v)(2u - v)^2 \, du \, dv == -52$

Note 13.5. (1) $\oint_{\partial S_1} F \cdot ds = \int_{t_0}^{t_1} F \circ C_1 \cdot C_1, dt = \int_0^2 (t^4, t^2 + t) \cdot (1,1) \, dt = \int_0^2 t^4 + t^2 + t \, dt = \frac{166}{15}$, (2) $\oint_{\partial S_2} F \cdot ds = \int_{t_0}^{t_1} F \circ C_2 \cdot C_2, dt = \int_0^1 [(2 + t)(2 + 2t)^3, (2 + 2t)^2 + (2 + t)] \cdot (1,2) \, dt = \int_0^1 (2 + t)(2 + 2t)^3 \, dt + \int_0^1 2[(2 + 2t)^2 + (2 + t)] \, dt = \frac{1549}{15}$, (3) $\oint_{\partial S_3} F \cdot ds = \int_{t_0}^{t_1} F \circ C_3 \cdot C_3, dt = \int_2^0 [(1 + t)(2 + t)^3, (2 + t)^2 + (1 + t)] \cdot (1,1) \, dt = \int_2^0 (1 + t)(2 + t)^3 + (2 + t)^2 + (1 + t) \, dt = -\frac{2416}{15}$, and (4) $\oint_{\partial S_4} F \cdot ds = \int_{t_0}^{t_1} F \circ C_4 \cdot C_4, dt = \int_1^0 (8t^4, 4t^2 + t) \cdot (1,2) \, dt = \int_1^0 8t^4 + 8t^2 + 2t \, dt = -\frac{79}{15}$. So $\oint_{\partial S_1} F \cdot ds + \oint_{\partial S_2} F \cdot ds + \oint_{\partial S_3} F \cdot ds + \oint_{\partial S_4} F \cdot ds = \frac{166}{15} + \frac{1549}{15} - \frac{2416}{15} - \frac{79}{15} = -\frac{780}{15} = -52$.

Note 13.6. $\nabla \times F = (\frac{\partial}{\partial x}, \frac{\partial}{\partial y}, \frac{\partial}{\partial z}) \times (xy^3, y^2 + x, 0) = (0, 0, 1 - 3xy^2) \Rightarrow (\nabla \times F) \circ T = (0, 0, 1 - 3xy^2) \circ T(u, v) = [0, 0, 1 - 3(u - v)(2u - v)^2] = (0, 0, 1 + 8u^2v - 5uv^2 - v^3 - 4u^3)$.

Note 13.7. $T_u(u, v) = (1, 2, 0)$, and $T_v(u, v) = (-1, -1, 0) \Rightarrow T_u \times T_v = (0, 0, 1)$.

13.4. GAUSS THEOREM

Definition 13.3. Let W be an oriented surface [Marsden and Tromba, 2011] Ch. 8 in a space with boundary denoted by ∂W, and F a C^1 vector field on W.

$$\oiint_{\partial W} F \cdot dw = \iiint_W (\nabla \cdot F) \cdot dW \tag{13.3}$$

Three transformations are required to simplify the integrals: $T_W : [\rho_0, \rho_1] \times [\theta_0, \theta_1] \times [\phi_0, \phi_1] \subset \mathbb{R}^3 \to W \subset \mathbb{R}^3$; $T_{\partial W} : [u_0, u_1] \times [v_0, v_1] \subset \mathbb{R}^2 \to \partial W \subset \mathbb{R}^3$

$\oiint_{\partial W} F \cdot dw = \int_{v_0}^{v_1} \int_{u_0}^{u_1} (F \circ T_{\partial W}) \cdot N \, du \, dv$, where $N : \mathbb{R}^2 \to \mathbb{R}^3, T_u \times T_v$; and

$\iiint_W (\nabla \cdot F) \cdot dw = \int_{\phi_0}^{\phi_1} \int_{\theta_0}^{\theta_1} \int_{\rho_0}^{\rho_1} (\nabla \cdot F) \circ T_W \;\cdot\; |J(T_W)| \, d\rho \, d\theta \, d\phi$.

Example 13.3. Let F vector field be $F(x, yz) = (x, 1, 1)$, and let W region be $W = \{(x, y, z) \mid x^2 + y^2 + z^2 \leq 1\}$. Verify the Gauss theorem.

Solution   **13.3.**   (i) $\iiint_W (\nabla \cdot F) \cdot dw = \int_{\phi_0}^{\phi_1} \int_{\theta_0}^{\theta_1} \int_{\rho_0}^{\rho_1} (\nabla \cdot F) \circ T_W$ ·

$|J(T_W)| d\rho \, d\theta \, d\phi = \int_0^\pi \int_0^{2\pi} \int_0^1 \rho^2 \sin\phi d\rho \, d\theta \, d\phi = \frac{1}{3} \int_0^\pi \int_0^{2\pi} \sin\phi \, d\theta \, d\phi =$

$\frac{2}{3}\pi \int_0^\pi \sin\phi \, d\phi = \frac{4}{3}\pi.$ (ii) $\oiint_{\partial w} F \cdot dw = \int_0^{2\pi} \int_0^\phi (F \circ T_{\partial W}) \cdot N \, d\phi \, d\theta =$

$\int_0^{2\pi} \int_0^\phi (\cos\theta\sin\phi, 1, 1) \cdot (-\sin^2\phi\cos\theta, -\sin^2\phi\sin\theta, -\sin\phi\cos\phi) \, d\phi \, d\theta =$

$\int_0^{2\pi} \int_0^\phi \sin^3\phi\cos^2\theta + \sin^2\phi\sin\theta + \sin\phi\cos\phi d\phi \, d\theta = \frac{4}{3}\pi$

Note 13.8. $T_W : [\rho_0, \rho_1] \times [\theta_0, \theta_1] \times [\phi_0, \phi_1] \subset \mathbb{R}^3 \to W \subset \mathbb{R}^3 = T_W : [0,1] \times [0,2\pi] \times [0,\pi] \subset \mathbb{R}^3 \to W \subset \mathbb{R}^3, (\rho\cos\theta\sin\phi, \rho\sin\theta\sin\phi, \rho\cos\phi) \Rightarrow J(T_W) = -\rho^2\sin\phi \Rightarrow |J(T_W)| = \rho^2\sin\phi.$

Note 13.9. $\nabla \cdot F = (\frac{\partial}{\partial x}, \frac{\partial}{\partial y}, \frac{\partial}{\partial z}) \cdot (x, 1, 1) = \frac{\partial}{\partial x}x + \frac{\partial}{\partial y}1 + \frac{\partial}{\partial z}1 = 1 \Rightarrow (\nabla \cdot F) \circ T_W = 1.$

Note 13.10. $T_{\partial W} : [\theta_0, \theta_1] \times [\phi_0, \phi_1] \to \partial W \subset \mathbb{R}^3, (\cos\theta\sin\phi, \sin\theta\sin\phi, \cos\phi \Rightarrow N = T_\theta \times T_\phi = (-\sin^2\phi\cos\theta, -\sin^2\phi\sin\theta, -\sin\phi\cos\phi).$

Note 13.11. $F \circ T_{\partial W} = (\cos\theta\sin\phi, 1, 1).$

Note   *13.12.* $\int_0^{2\pi} \int_0^\phi \sin^3\phi\cos^2\theta + \sin^2\phi\sin\theta + \sin\phi\cos\phi \, d\phi \, d\theta = \int_0^{2\pi} \cos^2\theta \, d\theta \int_0^\phi \sin^3\phi \, d\phi = \frac{4}{3}\pi$

The transformations mentioned above reduced the complexity of the integrals derived from these theorems.

REFERENCES

Abia, J. A. (2001). *Series de Fourier*. Universidad de Valladolid. http://www.ma.u va.es/antonio/ Teleco/Apun Mat2/Tema-13.pdf.

Beezer, R. (2017). *A first course of Linear Algebra*. University of Puget Sound, Tacoma, WA 98416, USA. http://linear.ups.edu/jsmath/0220/fcla-jsmath-2.20li6 1.html.

Bourbaki, N. (1998). *General Topology*. Springer-Verlag., New York, NY 10013, USA.

Bowers, N. L., Gerber, U., Hickman, C., Jones, D. A., and Nesbitt, C. J. (1997). *Actuarial Mathematics*. The Society Of Actuaries., Schaumburg, Illinois 60173, USA.

Brannon, R. (2002). Rotation: A review of useful theorems involving proper orthogonal matrices referenced to three-dimensional physical space. Sandia National Laboratories., Albuquerque, NM 87185-0820, USA.

Carrell, J. B. (2005). *Fundamentals of Linear Algebra*. University of British Columbia., Vancouver, BC V6T 1Z2, USA. https://www.math.ubc.ca/ carrel-l/NB.pdf.

Casella, G. and Berger, R. (2002). *Statistical Inference*. Duxbury Advanced Series, Boston, PWS-Kent, USA. https://fsalamri.files.wordpress.com/2015/02/casella berger statistical inference1.pdf.

CDC (2017a). Ebola outbreak in west africa. https://www.cdc.gov/vhf/ebola/outb reaks/2014-west-africa/index.html.

CDC (2017b). Ebola outbreak in west africa - case counts. https://www.cdc.gov/vhf/ebola/outbreaks/2014-west-africa/case-counts.html.

CDC (2017c). Ebola outbreak in west africa - outbreak distribution. https://www.c dc.gov/vhf/ebola/outbreaks/index.html.

Cherney, D., Denton, T., Thomas, R., and Waldron, A. (2013). *Linear Algebra*. Cre-ative Commons Attribution-NonCommercial-ShareAlike 3.0 Unported License, Davis, CA 95616, USA. https://www.math.ucdavis.edu/ linear/linear-guest.pdf.

Cornejo-Rodríguez, A. and Urcid-Serrano, G. (2005). *Optica Geometrica Resumen de Conceptos y Formulas. Parte I*. INAOE, Tonanzintla Puebla, 72840, México. http://wwwoptica.inaoep.mx/investigadores/urcidgesp/aca/og acorgurc2005.pdf.

Cullen, C. (2003). *Matrices and Linear Transformations*. Dover Publications, Mi-neola, NY 11501, USA.

Eddy, S. (1996a).Hidden markov models.*Curr Opin Struct Biol*, 3(6):361–365. DOI: 10.2174/1574893 610666151008012541.

Eddy, S. (1996b). Hidden markov models. *Current Opinion in Structural Biology*, 6(6):361–365.

Farmer, D., Schueller, A., and Guichard, D. (2012). *Elementary Calculus: An In-finitesimal Approach*. https://www.whitman.edu/mathematics/calculusonline/section16.02.htm.

Friedberg, S., Insel, A., and Spence, L. (1982). *Algebra Lineal*. Educal, México, CDMX 06040.

George, M. (1982). Affine Transformations, Transformation Geometry: An Introduction to Symmetry. Springer-Verlag., New York, NY 10013, USA.

Hardy, M. and Waters, H. (2009). *Actuarial Mathematics for Life Contingent Risks*. Cambridge University Press., New York, NY 10006, USA.

Hefferson, J. (2014). *Linear Algebra*. Saint Michael's College, Colchester, VT 05439, USA. http://joshua.smcvt.edu/linearalgebra/book.pdf.

Horowitz, L. (1996). *Emerging Viruses: AIDS & Ebola Nature, Accident or Intentional?* http://ethosworld. com/library/Leonard-G.-Horowitz   Emerging-Viruses-AIDS-%26-Ebola-Nature,-Accident-or-Intentional% 281996%29.pdf.

Jenkins, F. and White, H. (1937). *Fundamental of Optics*. McGraw-Hill, New York, NY 10020, USA.

Kardi, T. (2016). *Non Linear Transformation*. Revoledu Design. http://people.revoledu.com/kardi/tutorial/ Regression/nonlinear/NonLinearTransformation.htm.

Kolmogorov, A. and Fomin, S. (1970). *Introductory Real Analysis*. Dover Publica-tions, Englewood Cliffs, NY 07632, USA.

Malcev, A. (1963). *Foundations of Linear Algebra*. W H Freeman And Company, New York, NY 10004, USA.

Marsden, J. and Tromba, A. (2011). *Vector Calculus*. W H Freeman And Company, New York, NY 10004, USA.

Nathan, J. (2009). *Basic Algebra II*. Dover Publications, Mineola, NY 11501. Parkhomenk, P. and Parkhomenko, A. (1965). *Euclidean and Affine Transformations*. Academic Press., London, WC1B 3DP, U.K.

Pauling, L. (1995). *General Chemistry*. W H Freeman And Company, New York, NY 10004, USA.

Polanco, C. (2013). Selective antibacterial peptides: a review on their polarity. In Méndez-Vilas, A., editor, Microbial Pathogens and Strategies for Combating them: Science, Technology and Educational, chapter 4, pages 1307-1317. For-matex Research Center, Badajoz, SP 06002, Spain.

Polanco, C. (2014). Possible computational filter to detect proteins associated to influenza a subtype H1N1. *Acta Biochim Pol*, 61(4):693–698.

Polanco, C. (2015). Precision medicine and portable embedded system devices. (Letter to the editor) personalized medicine: Time for one-person trials. *Nature*, 520(6):609–611. DOI:10.1038/520609a.

Polanco, C. (2016a). Identification of antimicrobial peptides using eigenvectors. *Acta Biochim Pol*, 63(2):483–491. DOI: 10.18388/abp.2015˙993.

Polanco, C. (2016b). *Polarity Index in Proteins - A Bioinformatics Tool*. Bentham Science Publishers., Sharjah, 7917, UAE.

Polanco, C., Buhse, T., and Samaniego-Mendoza, J. L. (2015a). Bioinformatics tool to identify peptides associated to cancer cells. *Current Bioinformatics*, 10(5):632–638. DOI: 10.2174/1574893610666151008012541.

Polanco, C., Buhse, T., Samaniego-Mendoza, J. L., and Castañón-González, J. A. (2013a). Detection of selective antibacterial peptides by the polarity profile method. *Acta Biochim Pol*, 60(2):183–189.

Polanco, C., Buhse, T., Samaniego-Mendoza, J. L., and Castañón-González, J. A. (2013b). A toy model of prebiotic peptide evolution: the possible role of relative amino acid abundances. *Acta Biochim Pol*, 60(2):175–182.

Polanco, C., Buhse, T., Samaniego-Mendoza, J. L., Castañón-Gonzalez, J. A., and Arias-Estrada, M. (2014a). Computational model of abiogenic amino acid con-densation to obtain a polar amino acid profile. *Acta Biochim Pol*, 61(2):253–258.

Polanco, C. and Samaniego-Mendoza, J. L. (2009). Detection of selective cationic amphipatic antibacterial peptides by hidden markov models. *Acta Biochim Pol*, 56(1):167–176. DOI: 10.18388/abp.2015˙993.

Polanco, C., Samaniego-Mendoza, J. L., Buhse, T., Mosqueira, F. G., Negrón, A., Bernal, S. R., and Castañón-González, J. A. (2012). Characterization of selective antibacterial peptides by polarity index. *International Journal of Peptides*, 2012(4):54502. DOI: http://dx.doi.org/10.1155/2012/585027.

Polanco, C., Samaniego-Mendoza, J. L., Buhse, T., and Castañón-González, J.A. (2016a). Discrete dynamic system oriented on the formation of prebiotic dipeptides from Rode's experiment. *Acta Biochim Pol*, 61(4):717–726. DOI: 10.18388/abp.2016'1300.

Polanco, C., Samaniego-Mendoza, J. L., Buhse, T., Castañón-González, J. A., and Leopold-Sordo, M. (2014b). Polar characterization of antifungal peptides from APD2 database. *Cell Biochem Biophys*, 70(2):1479–1488. DOI: 10.1007/s12013-014-0085-3.

Polanco, C., Samaniego-Mendoza, J. L., Castañón-González, J. A., and Buhse, T. (2014c). Polar profile of antiviral peptides from avppred database. *Cell Biochem Biophys*, 70(2):1469–1477. DOI: 10.1007/ s12013-014-0084-4.

Polanco, C., Samaniego-Mendoza, J. L., Castañón-González, J. A., and Buhse, T. (2014d). (Letter to the editor) antimicrobial resistance: A global challenge. *Sci Transl Med*, 236(6):236ed10. DOI:10.1126/scitranslmed.3009315.

Polanco, C., Samaniego-Mendoza, J. L., Castañón-González, J. A., Buhse, T., and Estrada, M. A. (2014e). Computational model of abiogenic amino acid condensation to obtain a polar amino acid profile. *Acta Biochim Pol*, 61(2):253–258.

Polanco, C., Samaniego-Mendoza, J. L., Castañón-González, J. A., Buhse, T., and Sordo, M. (2013c). Characterization of a possible uptake mechanism of selective antibacterial peptides. *Acta Biochim Pol*, 60(4):629–633.

Polanco, C., Samaniego-Mendoza, J. L., Uversky, V. N., Castañón-González, J. A., Buhse, T., Sordo, M., Madero-Arteaga, A, Morales-Reyes, A., Tavera-Sierra, L., González-Bernal, J., and Estrada, M. A. (2015b). Identification of proteins associated with amyloidosis by Polarity Index Method. *Acta Biochim Pol*, 62(1):41–55.

Polanco, C., Castañón-González, J. A., Buhse, T., Samaniego-Mendoza, J. L., Arreguín-Nava, R., and Villanueva-Martínez, S. (2013d). Índice de Saturación Modificado en el Servicio de Urgencias Medicas. Gac Med Mex, (149): 417-424. DOI: http://dx.doi.org/10.1155/2013/ 213206.

Polanco, C., Castañón-González, J. A., Buhse, T., Uversky, V. N., and Amkie, R. Z. (2016b). Classifying lipoproteins based on their polar profiles. *Acta Biochim Pol*, 63(2):235–241. DOI: 10.18388/abp.2014-918.

Polanco, C., Castañón-González, J. A., Macías, A., Samaniego-Mendoza, J. L., Buhse, T., and Villanueva-Martínez, S. (2013e). Detection of severe respiratory disease epidemic outbreaks by cusum-based overcrowd-severe-respiratory-disease-index model. Computational and Mathematical Methods in Medicine, (213206). DOI: http://dx.doi.org/10.1155/2013/ 213206.

Polanco, C., Castañón-González, J. A., and Uversky, V. N. (2014f). (Letter to the editor) protein misfolding, congophilia, oligomerization, and defec-tive amyloid processing in preeclampsia. *Sci Transl Med*, 16(6):245ra92. DOI:10.1126/scitranslmed.3008808.

Polanco, C., Castañón-González, J.A., Uversky, V. N., Buhse, T., Samaniego-Mendoza, J. L., and Calva, J. J. (2016c). Electronegativity and intrinsic disorder of preeclampsia-related proteins. Acta Biochim Pol, 64(1): 99-111. DOI: 10.18388/abp.2016 1300.

Rincón, L. (2014). *Introducción a la probabilidad*. Las prensas de Ciencias, México City, 04320, México.

Rojo, A. (1973). Álgebra II. El Ateneo, Buenos Aires, AR 3100.

Rorres, C. and Anton, H. (2011). Limusa Noriega, *Introducción al Álgebra Lineal* México City, 06000, México.https://bibliotecavirtualmatematicasunicaes.files.wordpress.com/2011/1/intrioduccic3b3n-al-c3a1lgebra-lineal-3ra-edicic3b3n-ho ward-anton1.pdf.

Rotman, J. (2002). *Advanced Modern Algebra*. Pearson Education, Upper Saddle River, NJ 07458, USA.

Rudin, W. (1964). *Principles of Mathematical Analysis*. McGraw-Hill, New York, NY 10020, USA. https://notendur.hi.is/vae11/%C3%9Eekking/principles of ma thematical analysis walter rudin.pdf.

Sheldon, R. (1976). *A first course in probability*. Prentice Hall, Upper Saddle River, New Jersey 07458, USA. http://zalsiary.kau.edu.sa/Files/0009120/Files/119387 A First Course in Probability 8th Edition.pdf.

Simanek, D. (2017). Graphs. https://www.lhup.edu/ dsimanek/scenario/errorman/-graphs2.htm.

Spiegel, M. R. (2001). *Statistics*. McGraw-Hill, New York, NY 10020, USA. Stewart, J. (1998). *Calculus: Early Transcendentals*. Cengage Learning., Boston, MA 02210, USA.

Thompson, K. (1996). *Calculus*. Department of Mathematics, Corvallis, OR 97331-4605, USA. https://math.oregonstate.edu/home/programs/undergrad/CalculusQu estStudyGuides/vcalc/lindep/lindep.html.

Wang, G. and Wang, Z. (2009). APD2: the updated antimicrobial peptide database and its application in peptide design. *Nucleic Acids Res*, 37(37):D933–D937. DOI: 10.1093/nar/gkn823.

Wikipedia (2017a). Line integral. https://en.wikipedia.org/w/index.php?title=Lineintegral&oldid= 763594821.

Wikipedia (2017b). Linear algebra. https://en.wikipedia.org/w/index.php?title=Basi s (linear algebra) &oldid=759 297017.

Wikipedia (2017c). Linear span. https://en.wikipedia.org/w/index.php?title=Linea r span&oldid=761783161.

Wikipedia (2017d). Logarithm. https://en.wikipedia.org/w/index.php?title=Logarit hm&oldid=784440999.

Wikipedia (2017e). Shear mapping. https://en.wikipedia.org/w/index.php?title=Shear mapping&oldid= 751000234.

SUBJECT INDEX

A

Acoustics 41, 43, 65
Affine transformation 1, 2, 4
Albino 70, 71
Algebraic structure 9
Algorithms 45, 50, 67, 68
Amino acids 50, 51, 52, 53, 56
Artificial neural network 65
Automorphism 2, 4
Auxiliary 31, 34, 35

B

Bayes theorem 70, 71, 72
Binomial distribution 82
Biometrics model 46

C

Cancer cells 51, 52, 53
Classical probability definition 69, 71
Complexity 1, 90
Composition transformations 2, 57, 60
Continuous function 2
Continuous real valued functions 5, 9
Cost of capital 45, 46
Cybernetics 65
Cylindrical transformation 62

D

Deterministic models 67
Double homothetic factor 30

E

Ebola outbreak 67, 68
Eigenvalues 72, 74, 78
Eigenvectors 72, 74, 78
Electronic devices 67

Element, unique 2, 4
Endomorphism 2, 3, 4, 58, 60, 62
Equivalence rules 50
Exponential distribution 82, 83
Exponential function 38, 39

F

First descendant 70, 71
Fourier transform 65
Function 2, 4, 9, 13, 22, 30, 32, 33, 34, 38, 39,
   41, 43, 44, 46, 47, 48, 53, 59, 62, 69, 80,
   81, 82 83, 85, 86
  density 82, 85
  linear 30, 47
  real 41, 62
Function space 8, 13

G

Gauss theorem 57, 87, 89
General format 61
Genes 69, 70, 71, 72
  following combination of 71
  pair of 71
Genotypes 72, 73
Green theorems 57, 87, 88

H

Hidden Markov models 67
Homothetic transformation multiplies 7
Homothetic transformations 7

I

Incident beam 80
Integration area 57, 60
Invariant subspace 29
Inverse transformation 2, 4, 31, 32, 33, 34, 35
Isomorphism 2, 3, 4

J

Jacobian determinant 57, 62

L

Laplace transform 65, 66
Least squares method 30, 35, 36
Linear 1, 2, 3, 4, 5, 6, 7, 8, 9, 10, 13, 14, 17
    22, 24, 25, 26, 27, 28, 29, 30, 32, 33, 34,
    38, 42, 45, 46, 47, 48, 49, 50, 58, 60, 65,
    67, 78, 80, 81, 85, 86
   dependence 27, 28
   epresentation 50
   Independence 27
   programming 45, 47
   space 2, 3, 4, 7, 8, 9, 10, 13, 14, 17, 22, 24,
    25, 26, 27, 28, 30
   transformation 1, 2, 3, 4, 5, 6, 8, 29, 30, 32,
    33, 34, 38, 42, 45, 46, 48, 49, 58, 60, 65,
    67, 78, 80, 81, 85, 86
Logarithmic 32, 33, 34, 38, 39, 40
   functions 38
   graph paper 32, 33, 34
   model 30, 31
   scale 39, 40

M

Mapping, nonlinear 1
Markov model 53, 68
Matrix 8, 17, 26, 57
   derivatives 57
   space 8, 17, 26
Maximum profit 48, 49
Mortality rate 45, 46

N

Nonlinear equation 30, 31, 34, 35, 36
Normal distribution parameters 85
Nucleotides 50

O

One-element 26
Operator algebra 1
Oriented surfaces 61, 87, 88, 89
Overcrowd severe respiratory disease index
   (OSRDI) 68

P

Paralelogram 60
Pathogen agents 51
Pathogenic action 51
Pearson correlation coefficient 36, 37
Plane, coordinate 60, 61
Plants type 73
Polarity index method 50
Polar profile 50, 51, 53, 55, 56
Polar transformation 60
Polyester 48
Polynomial space 8, 14
Probability, temporal 46
Probability function 70
Processing, natural language 65
Properties, spanning 27
Protein groups 51, 50, 51, 53
   representative 50, 51
Proteins, target 50, 51
Protein structure 50, 52

R

Random variable 46, 83, 84, 85
Real time monitoring 67, 68
Reciprocal model 30, 34
Reflection transformations 2, 6, 29, 57, 60
Rotation angle 5, 6
Rotation transformations 2, 5, 29, 57, 60

S

SCAAP group 53, 55

www.ingramcontent.com/pod-product-compliance
Lightning Source LLC
Chambersburg PA
CBHW041729210326
41598CB00008B/829

* 9 7 8 1 6 8 1 0 8 7 1 2 2 *